Electrical Drive Simulation with MATLAB®/Simulink®

The chapters of this book discuss the modeling of electric drives, taking into account their relationship with the technological process they serve, which significantly affects the composition, layout, and characteristics of the electric drive. There are no published books of this kind, and this book fills a gap in the literature. This book deals with electric drives for rolling mills, paper machines, a number of several hoisting and transport devices; these installations are very common and very complex, so that modeling methods in their development and study are mandatory.

The book focuses on issues such as the transmission of torque by elastic shafts, the transmission of torque by an endless elastic belt in paper machines and conveyors, the transmission of torque by friction of pressed rolls in the paper industry, the consideration of the elastic properties of long ropes in some hoisting and transport machines, and the effect of swinging a moving load in such machines. More than 100 models of the electrical drives that are made with the use of the program environment MATLAB®/Simulink® are appended to this book. The aims of these models are to aid students studying electrical drives of the various manufacturing machines, to facilitate the understanding of various electrical drive functions, and to create a platform for the development of systems by readers in their fields.

This book can be used by engineers and investigators as well as undergraduate and graduate students to develop new electrical drives and investigate the existing ones.

Electrical Drive Simulation with MATLAB®/Simulink®

Selected Technologies

Viktor Perelmuter

CRC Press

Taylor & Francis Group

Boca Raton London New York

CRC Press is an imprint of the
Taylor & Francis Group, an **informa** business

Cover image: Shutterstock

First edition published [2024]
by CRC Press
6000 Broken Sound Parkway NW, Suite 300, Boca Raton, FL 33487-2742

and by CRC Press
4 Park Square, Milton Park, Abingdon, Oxon, OX14 4RN

CRC Press is an imprint of Taylor & Francis Group, LLC

ISBN: 9781032495552 (hbk)
ISBN: 9781032495576 (pbk)
ISBN: 9781003394419 (ebk)

DOI: 10.1201/9781003394419

Typeset in Times
by codeMantra

Access the Support Material: *www.routledge.com/9781032495552*

To my colleagues—electric drive specialists,
living and already gone from us

Contents

Preface

A significant part of the electrical energy produced is used by converting it into mechanical energy using electric motors. The complex, which contains all the devices necessary for such a conversion, is called an electric drive. An integral part of the electric drive is the operating member, which connects the rotational movement of the motor shaft with the working movement of the process unit. It is critically important that the composition, characteristics, and parameters of the electric drive to the greatest extent correspond to the requirements that demand the optimal operation of the technological installation, the technological process, in which this electric drive is used. Modern electric drives of many production processes and mechanisms are very complex devices, and the availability of modern and effective methods of mathematical modeling of such systems can significantly facilitate and speed up the creation of such electric drives. The graphical programming language MATLAB®/Simulink® and the set of Electrical/Specialized Power Systems blocks included in it greatly facilitate the creation of models of electrical engineering objects.

There are a number of very interesting and useful books that describe the design, systems of power supply and control of modern electric drives, and methods for their modeling. However, these books do not discuss how the characteristics of these drives relate to the process which these drives serve, and this book aims to fill that gap. Of course, there is a wide variety of technological processes and operations that make variety of demands on electric drives. The author chose those that are very common and complex enough that it is difficult to do without simulation when developing them. In addition, their choice was influenced by the fact that the author was involved in the development of most of these electric drives during his 40 years of engineering activity.

The first chapter deals with the modeling of reversible and continuous hot and cold rolling mills. When modeling reversible hot rolling mills, the main attention is paid to the elasticity of the shafts that transmit rotation from the motor to the work rolls, as well as the ratio of the speeds of the work rolls and the rolling stand as a whole, providing the desired shape of the processed ingot and the maximum productivity of the mill. Simultaneous simulation of the operation of the main electric drive of the stand and its pressure screws, which set the distance between the rolls (roll clearance) and significantly affect the performance of the mill, makes it possible to visualize and evaluate the rolling process on mills of this type.

Plate mills belong to the rolling mills of the same type. Currently, old mills are being modernized and new ones are being built due to the increasing demand for wide hot-rolled sheets for nuclear power engineering, for the construction of large tankers, for the production of large diameter pipes, and so on. New plate mills are characterized by an increased length of the roll barrel—up to 5–5.5 m and the great motor power. This book simulates a real rolling mill of this type with a stand motor power of $2 \times 12,000$ kW. At the same time, the operation of the pressure screws and the process of changing and adjusting the strip thickness using a hydraulic device for moving the lower roll are simulated.

Continuous hot rolling mills consist of roughing and finishing groups. In modern mills, rolling in the roughing group occurs simultaneously in several stands, and the problem arises of regulating their speeds in such a way as to avoid strip tension in the absence of tension sensors. Ways to solve this problem are modeled in this book. In the finishing group, rolling occurs simultaneously in all stands. To maintain the tension between them, loopers are installed that have a non-linear characteristic. This book simulates the rolling process in the finishing group, including the regulation of interstand tensions with the help of the loopers, as well as the processes of regulating the thickness of the rolled strip. In the same chapter, the operation of flying shears is simulated, designed to cut off the front, defective end of the strip and to cut the strip into cut lengths.

When modeling cold strip rolling mills, first of all, the modeling of a coiler operating in various modes, with a direct or indirect tension controller, is considered. When modeling a reversing cold rolling mill, the joint operation of the drives of the stand, winder and unwinder, is simulated, with control of the thickness of the rolled strip and taking into account the speed control laws of the mill, ensuring accurate stopping and fast reverse. When modeling continuous cold rolling mills, the joint work of all stands is considered, including the regulation of interstand tensions and the regulation of strip thickness.

The second chapter of this book is devoted to modeling electric drives of paper-making machines (PM). A feature of such electric drives is the fact that the moment of inertia of the load significantly exceeds the moment of inertia of the motor, so that the consideration of torsional vibrations of the shafts is mandatory. This book sequentially discusses the modeling of electric drives of individual sections of the PM: forming, press, drying, calender, and reel, in each of which the simulation of electric drives has its own characteristics. In the forming section, this is the transfer of torque to a moving infinite grid by friction. In the press section, this is a simulation of the distribution of the load between rotating driven rolls pressed against each other with varying force. When modeling the drying section, the need for synchronous rotation of the shafts of adjacent groups of the section connected by a paper web is taken into account, with a small speed difference caused by the drying of this web and taking into account the changing amount of condensate in the drums. The models of the drives of the peripheral reels developed in this book make it possible to simulate various modes, including speed control, tension control, paper web breakage, and so on. This chapter also discusses the simulation of a supercalender and a double-drum slitter.

The third chapter discusses the modeling of some types of lifting and transport machines. When modeling a traveler crane, special attention is paid to modeling swinging of a suspended load and ways to eliminate it. When modeling the cable-crane electric drive, the elastic connection between the winch drum and the trolley through a long cable is taken into account. When modeling the electric drive of the mine hoist, the elastic properties of the ropes are taken into account, which, with their large length, have a significant impact on the operation of the mine hoist. There is a wide variety of designs and layouts of conveyors of great length (up to 1 km or more), but only three of them are modeled in this book. To take into account the elasticity of the conveyor belt, the finite element method is used. A long conveyor is also modeled, in which, in addition to the head ones, additional pulleys with an electric drive are installed along the conveyor line; control of the load ratio of electric drives depending on the load of the conveyor is modeled.

More than 100 models of electric drives of the types mentioned above are included with this book. These models are found at www.crcpress.com/9781032495552. The reader has to put all the appended files in the same folder in the MATLAB catalog; this folder has to be defined as the starting one.

During the work with the aforementioned models, the reader can investigate them with the given or the chosen parameters, use them as a basis for the analogous models developed by him or her, and so on. Interactive work with a computer is supposed to be done while reading the book.

By reading this book and working with models, keep the following in mind. For almost every drive application in this book, power calculations are performed, sometimes using simplified methods. This not only allows to choose the right motor but also makes it possible to understand better the features and modes of operation of the electric drive in this application. When creating models, optimal decisions were not always made, since the goal was to show the variety of possible approaches. In this regard, the issues of unification that are very important in practical work were not taken into account. For power supply and control of electric motors, mainly typical systems from the set of Electrical/Specialized Power Systems blocks are used, perhaps with some changes, or previously developed by the author and included in books published by CRC Press.

With the employment of the set of Electrical/Specialized Power Systems blocks, a very important problem is the compatibility of the developed models with diverse versions of this set, bearing in mind that every year two versions appear; at this, the new blocks appear, some blocks are declared obsolete, and some of these versions have new features that influence the model performance. Given the possible contingent of readers, it is unrealistic to assume that all of them have access to the latest versions. The function of the appended models was checked with the versions R2019b and R2023b. It can be assumed that they will function with the intermediate versions.

This book can be used by engineers and investigators in the development of new electrical systems and investigations of the existing ones. It is very useful for students of higher educational institutions during the study of electrical fields, in graduation work and undergraduate's thesis.

MATLAB® is a registered trademark of The Math Works, Inc. For product information, please contact:
  The Math Works, Inc.
  3 Apple Hill Drive
  Natick, MA 01760-2098
  Tel: 508-647-7000
  Fax: 508-647-7001
  E-mail: info@mathworks.com
  Web: http://www.mathworks.com

Author

Viktor Perelmuter, DSc, earned a diploma in electrical engineering with honors from the National Technical University (Kharkov Polytechnic Institute) in 1958. Dr. Perelmuter earned a candidate degree in technical sciences (PhD) from the Electromechanical Institute Moscow, Soviet Union, in 1967, a senior scientific worker diploma (confirmation of the Supreme Promoting Committee by the Union of Soviet Socialist Republics Council of Ministers) in 1981, and a doctorate degree in technical sciences from the Electrical Power Institute Moscow/SU in 1991.

From 1958 to 2000, he worked in the Research Electrotechnical Institute, Kharkov, Ukraine, in the thyristor drive department, during which he also served as Department Chief (1988–2000). During this time, he developed DC electric drives for rolling mills of various types, including such a specific mill as a planetary mill for rolling sheet metal. Starting with the employment of the rotating electrical machines, magnetic and thyristor amplifiers as the exciters, with the advent of powerful thyristors, he carried out the development of thyristor electric drives for reversible and continuous cold and hot rolling mills. He was the head of work on the creation in the USSR of a series of complete thyristor electric drives for rolling mills with a power of up to 12,000 kW. He repeatedly took part in putting power electric drives into operation at metallurgical plants. This work was commended with a number of honorary diplomas. He was also involved as a consultant on problems related to paper-making machines, mine hoists, and conveyors.

Between the years 1993 and 2000, he was the director of the joint venture Elpriv, Kharkov. During 1965–1998, he was the supervisor of the graduation works at the Technical University, Kharkov. Dr. Perelmuter was a chairman of the State Examination Committee in the Ukrainian Correspondence Polytechnic Institute from 1975 to 1985. Simultaneously with his ongoing engineering activity, he led the scientific work in the fields of electrical drives, power electronics, and control systems.

He is the author or a coauthor of 12 books and approximately 75 articles and holds 19 patents in the Soviet Union and Ukraine. Since 2001, up to Ukraine–Russia war, Dr. Perelmuter had been working as a scientific advisor in the National Technical University (Kharkov Polytechnic Institute) and in Ltd "Jugelectroproject", Kharkov. He is also a life member of the Institute of Electrical and Electronics Engineers.

1 Simulation of the Electrical Drives of Rolling Mills

1.1 ELECTRIC DRIVES OF REVERSING HOT ROLLING MILLS

Rolling mills of this type include blooming, slabbing, and plate mills. They produce work pieces for the subsequent production of final products: blooms—square-section blanks for the production of long products, pipes, rails, and so on; slabs—flat blanks for the production of sheet products; plate mills—hot-rolled sheets used in nuclear power, in the production of large diameter pipes, in shipbuilding, and so on. A feature of these mills is the reverse mode of operation: in the reversible stand of the mill, rolling occurs in both directions and takes several passes. The number of passes is odd, since the ingots enter one side of the stand and are given out from the other. After each pass, the motor reverses. If a blooming has one stand with horizontal rolls, then a slabbing has two stands—horizontal and vertical, in which simultaneous rolling takes place. A plate mill has one, two, or three stands, in the latter case one of them with vertical rolls.

The complex of a reversible hot rolling mill, in addition to the main electric drives of the stands, includes a number of other mechanisms, of which only pressure screws (or screw-downs) will be considered, which serve to set the gap between the rolls and largely determine the productivity of the entire mill.

1.1.1 BLOOMING ELECTRIC DRIVES

The number of passes during bloom rolling can be 13–21 or more. The rolls capture the ingot at a relatively low speed (20%–30% of the maximum speed in this pass), which is explained by the conditions for a reliable, shockless grip. After the capture, the mill accelerates with the metal in the rolls, and before the end of rolling, the speed is reduced so that the ingot is not thrown a long distance from the stand.

A feature of blooming rolling is that after several ordinary passes the ingot is turned over, that is, rotates 90°. If turning over is due after the next pass, the ejection speed is not reduced so that the ingot can reach the manipulator. The speed also does not decrease in the last pass.

The reduction in the pass $\Delta h = h_0 - h_1$ (Figure 1.1) is determined by the allowable rolling force and motor power, as well as the maximum possible angle of bite α, which, when rolling on blooming, is $20° - 25°$, so that

$$\Delta h_{max} = R\alpha^2. \tag{1.1}$$

DOI: 10.1201/9781003394419-1

1

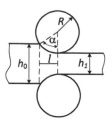

FIGURE 1.1   Scheme of rolling deformation.

There are a number of formulas for determining the rolling force for given values of h_0 and h_1. In general, these formulas consist of two factors, one of which is equal to the yield strength for the given material under given conditions σ_s, and the second is determined by the geometry of the compression zone and the friction coefficient between the metal and the rolls f.

There are a number of curves describing the dependence of σ_s on the percent reduction $\varepsilon = \Delta h / h_0$, temperature $T°C$, strain rate $u = V\varepsilon / l$, where V is the speed of metal exit from the rolls, mm/s, l is the horizontal projection of the contact arc (see Figure 1.1), $l = \sqrt{R\Delta h}$, mm. However, it is inconvenient to use these curves under simulation, and it is preferable to use analytical expressions, numerous examples of which are given in Ref. [1]. Further, we will use relation (1.2) for the steel that contains carbon 0.45% (analog C45—DIN, 1.0503—EN, M1044—USA):

$$\sigma_s = \frac{1330\varepsilon^{0.252} u^{0.143}}{e^{0.0025T}} \ \text{N/mm}^2. \tag{1.2}$$

As for the second factor, to calculate the average unit roll force p_a, we use the simple formulas given in Ref. [2]:

$$p_a = 1.15\sigma_s n_1 n_2 \tag{1.3}$$

$$n_1 = 1 + \frac{fl}{h_0 + h_1} \tag{1.4}$$

$$n_2 = \left(\frac{h_0 + h_1}{2l} \right)^{0.3}. \tag{1.5}$$

The coefficient of friction f is determined by the formula

$$f = k_f (1.05 - 0.0005T) \tag{1.6}$$

where $k_f = 1$ during an ingot bite at a speed of up to 1.5 m/s and $k_f = 0.8$ during rolling. The total pressure of a metal on the rolls (rolling force) is:

$$P = p_a lb \tag{1.7}$$

where b is the mean width of the ingot, $b = 0.5(b_0 + b_1)$, b_0 is the width of the ingot before the pass, and b_1 is after the pass. The spread $\Delta b = b_1 - b_0$ can be calculated as follows:

If $h_0/b_0 < 1$, then

$$\Delta b = 1.15 \frac{\Delta h}{2h_0}\left(l - \frac{\Delta h}{2f}\right) = \Delta b_1 \tag{1.8}$$

If $h_0/b_0 > 1$, then

$$\Delta b = \Delta b_1\left(1 + \frac{h_1 + 0.5\Delta h - fl}{2R}\right) \tag{1.9}$$

The rolling torque T_r is

$$T_r = 2P\psi l \tag{1.10}$$

where ψ is the coefficient of the arm of force, then $\psi = 0.5$ is taken.

The above calculations are implemented in the *Blm1.m* program and in the **Rolling** subsystem (**Rev_Torque** model). This model makes it possible to determine the forces and torques of rolling through the passes. The diameter of the rolls is set in the subsystem dialog box. The program is considered, in which the ingots with initial dimensions of 960×825 mm are rolled in blooms 350×350 mm in 13 passes, Ref. [3], Table 1.1. The letter K indicates turning over after this pass. As can be seen from the table, very large reductions are realized in some passes.

TABLE 1.1
Draughting Schedule 960 × 825 mm in 350 × 350 mm in 13 Passes

Pass Number	Size before Pass, $h \times b$	Size after Pass, $h \times b$
1	960 × 825	840 × 825
2 K	840 × 825	740 × 830
3	830 × 740	720 × 755
4	720 × 755	620 × 770
5	620 × 770	530 × 780
6 K	530 × 780	440 × 790
7	790 × 440	700 × 455
8	700 × 455	610 × 470
9	610 × 470	530 × 480
10 K	530 × 480	450 × 490
11	490 × 450	400 × 460
12 K	400 × 460	325 × 470
13	470 × 325	350 × 350

The initial temperature is assumed to be 1250°C with a decrease of 10°C after each pass. Figure 1.2 shows the changes in the dimensions of the ingot and the torque of rolling. It can be seen that the stand motors must briefly allow a total torque of up to 4 MN-m.

In the real blooming, an individual electric drive of rolls was used with DC motors with a power of 6800 kW each, 60/90 rpm, overload 2.25 for 15 s, that is, the maximum rolling torque for a short time may be $T_{r\ max} = \dfrac{2 \times 6800 \times 30 \times 2.25}{\pi \times 60} = 4.9$ MN-m. Moment of inertia of the motor is 75×10^3 kg-m² and must be increased by ~10% to account for the attached rotating parts.

In the model **Blum1**, the motor is simulated by a simplified way by the equation

$$J\frac{d\omega r}{dt} = T_m - T_r \qquad (1.11)$$

In turn, the motor torque T_m is determined by the output of the controller of the motor rotational speed ω_r, and the lag of the control system is modeled by a first-order element with a time constant of 8 ms. A generalized motor with a rated torque of 2160 kN-m and a rated rotational speed of $60 \times \pi/30 = 2\pi$ rad/s is assumed in the model. The parameters of the ingot along the passes and the speed of rotation of the

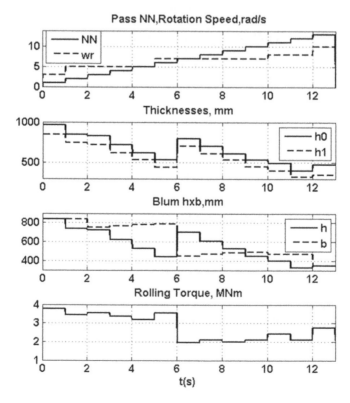

FIGURE 1.2   Bloom rolling process 960 × 825 mm in 350 × 350 mm.

rolls (rad/s) are set in the **Rolling_Parameters** subsystem. The motor acceleration is taken equal to $a_a = 4.5\,\text{rad/s}^2$ in the first six passes and $a_a = 7\,\text{rad/s}^2$ in the subsequent ones, and the deceleration is $a_d = -7\,\text{rad/s}^2$ in all passes. Appropriate operations are performed in the **Rate Limiter** subsystem, the scheme of which is shown in Figure 1.3. With signal ACC = 0 $a_a = 4.5\,\text{rad/s}^2$, and with ACC = 1 $a_a = 7\,\text{rad/s}^2$, the device functions correctly in reversible electric drives.

Rolling of the ingot weighing 150 kN is simulated, the specific weight of which at the assumed temperature is assumed to be 76 kN/m³, so that its length after the passage i is equal to $L_i = \dfrac{150 \times 10^6}{76 \times h_i \times b_i}\ m$ and reaches 16.1 m with the initial length of 2.5 m.

After the end of the running pass, 1 is added to its number, and the values for setting the speed and rolling torque in the next pass are formed in the **Rolling Parameters** subsystem. Using *Rem modulo 2* function, it is determined whether the pass number is even or odd, and depending on this, the sign of the speed reference is set.

The control over the passage of the ingot is carried out in the **Plate_Movement** subsystem. The length of the ingot in this pass L_i is compared with the already rolled segment L_t, which is calculated as $L_t = R \displaystyle\int_{t_0}^{t} \omega_r dt$, where t_0 is the moment of ingot bite. It is assumed that this occurs when the rotation speed after the motor reverses after the end of the previous pass reaches 2 rad/s. The braking distance at the initial speed of ω_0 is equal to $S_t = \dfrac{R\omega_0^2}{2a}$. When $L - L_t = S_t - b$, a signal is generated that causes the motor to enter deceleration mode. Here b is a parameter that determines the speed of the ingot ejection. When L_t becomes equal to L, the end-of-pass signal is generated (the **Bestable trigger** is reset), and the load torque becomes equal to the value of the idle torque (assumed to be 5%). As already mentioned, the trigger is set when the rotation speed after the motor reverses reaches 2 rad/s.

The model takes into account the need to increase the ejection speed during passes followed by turning over and in the last pass. For this purpose, in these passages, the start of braking occurs not under the condition $L - L_t = S_t - b$, but at $L = L_t$, that is, at the time of ingot exit from the rolls. The corresponding switches are carried out in the **Canting_Passes** subsystem.

The process is shown in Figure 1.4. If, when considering on a large scale, we determine the speed of rotation of the rolls at the moments when the ingot leaves the stand V_{out}, then we obtain the results shown in Table 1.2. It can be seen that in reality, after passes 2, 6, 10, 12, and 13, the exit speeds are equal to the rolling

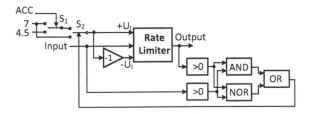

FIGURE 1.3 Diagram of the rate limiter for blooming drive.

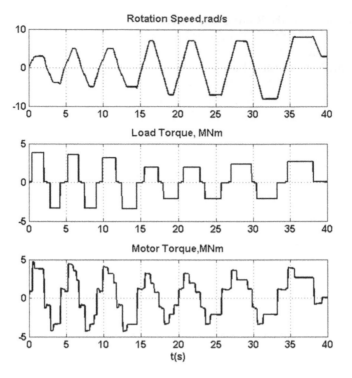

FIGURE 1.4   Blooming rolling cycle.

TABLE 1.2

Velocities, When the Ingot Leaves the Stand, rad/s

N	1	2	3	4	5	6	7	8	9	10	11	12	13
V_r	3	4	5	5	5	5	7	7	7	7	7	8	8
V_{out}	2.5	4	2.5	2	2.5	5	2.6	2.5	2.5	7	2.5	8	8

speed V_r. Motor power (**Power** oscilloscope), reaching 21 MW in steady-state conditions, increases to 30 MW during acceleration. The motor current value in pu is estimated as the value of the torque multiplied by the degree of field weakening of the motor. It can be seen that this value does not reach the tripping value equal to 2.5. If we measure the root mean square value of the current during the rolling of one ingot for 40 s, we will obtain a value of 1.11, without taking into account the pause between the ingots.

The rolling stand, together with the motor, is a complex mechanical structure, the dynamics of which can be modeled as a set of several concentrated masses connected by elastic shafts. An example of such a model is given in Ref. [4]. To build such a model, it is necessary to have detailed information about the structural elements of the stand. To take into account the elasticity of the shafts, the two-mass model is often considered, which is described by the equations

$$T_e = C \int (\omega_1 - \omega_2) dt + B(\omega_1 - \omega_2) \tag{1.12}$$

T_e—the torque transmitted from the motor to the rolls through the elastic axis, ω_1 is the motor rotation speed, ω_2 is the roll rotation speed, C and B are coefficients. Further, we set $B = 0$. These speeds can be found as

$$\frac{d\omega_1}{dt} = \frac{T_m - T_e}{J_1} \tag{1.13}$$

$$\frac{d\omega_2}{dt} = \frac{T_e - T_r}{J_2} \tag{1.14}$$

T_m is the motor torque; J_1 is its moment of inertia; T_r is the rolling torque; J_2 is the moment of inertia of the roll with the adjoined details. For such a system, the moment of inertia of the rolls is much less than the motor torque [4]. It is taken for the considered model $J_1 = 75{,}000 \text{ kg/m}^2$, $J_2 = 7500 \text{ kg/m}^2$. As for the stiffness coefficient C, this estimate can be done by the next way.

The diameter of the roll that transfers the rolling torque is defined from the following expression:

$$D > \sqrt[3]{\frac{16 T_{cr}}{\pi \tau \times 10^6}} \tag{1.15}$$

where T_{cr} is the limiting value of the transferred torque, N-m, τ is the allowable tangential stress that is taken usually 80–100 N/mm². With known D

$$C = \frac{\pi \tau D^4 \times 10^9}{32 l} \tag{1.16}$$

where l is the shaft length, m. For the blooming under consideration, the maximal torque of one roll is $T_{max} = 2.45$ MN-m; taking $T_{cr} = 2 T_{max}$ and $\tau = 80$ N/mm², we receive

$$D > \sqrt[3]{\frac{16 \times 2 \times 2.45 \times 10^6}{\pi \times 80 \times 10^6}} = 0.68 \text{ m}.$$

It is taken $D = 0.75$ m. In the modeled rolling mill, the top motor is located closer to the mill (Figure 1.5a), so we take $l = 8$ m for the top roll and $l = 16$ m for the bottom roll, so that $C_{top} = \dfrac{\pi \times 80 \times 0.75^4 \times 10^9}{32 \times 8} = 310$ MN-m/rad and $C_{bot} = \dfrac{\pi \times 80 \times 0.75^4 \times 10^9}{32 \times 16} = 155$ MN-m/rad.

In models **Blum2, 2a, and 2b**, an individual drive of blooming rolls with DC motors is simulated. The motors have a power of 6800 kW at a voltage of 930 V and a rated current of 7700 A. The rotational speed is 60/90 rpm, which corresponds to

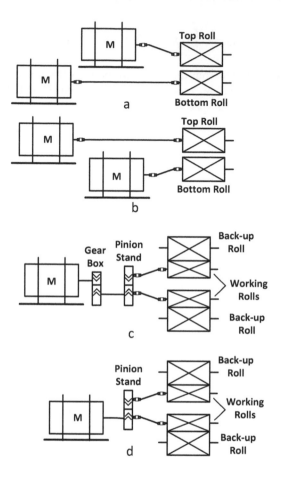

FIGURE 1.5 Possible designs of the rolling stand (a) Individual gearless roll drive, top motor closer; (b) Individual gearless roll drive, bottom motor closer; (c) Group drive of the four-high mill stand with gear; (d) Group gearless drive of the four-high mill stand.

6.28/9.4 rad/s. The moment of inertia of the motor is 75,000 kg-m², and the inertia of the connected rotating parts is 10 times less, the rated torque is 1082 kN-m. The resistance and inductance of the armature are 6 mΩ and 0.18 mH, respectively. The rated excitation flux is 0.284 Wb. Coefficient $L_{af} = (930 - 7700 \times 6 \times 10^{-3})/(0.284 \times 6.28) = 496$. The excitation winding resistance is 4.9 Ω, and the rated excitation current is 40 A. The parameters of the magnetization curve are given in Table 1.3.

TABLE 1.3

Dependence of the Flux Ψ on the Excitation Current I_f.

I_f, A	0	10	20	30	40	50
Ψ, Wb	0	0.1	0.18	0.24	0.284	0.32

The motors are powered by reversible thyristor converters without circulating currents. The selection of the active thyristor bridge is carried out by the logic switching unit LSU depending on the combination of the signs of the reference signals and the actual value of the motor current [5]. The LSU scheme is given in Figure 1.6.

The signal of the current direction (output of the speed controller) comes to the input **SIGN I***.

The absolute value of the load current comes to the input |**I**|. If the current value is more than 1%–2% of the rated value (holding current), the logical 0 at the **Relay2** output does not permit for the signal **SIGN I*** to change the LSU output state. When the current value is small, the logical 1 appears at the output of one of the OR gates (depending on the **SIGN I*** polarity), and the flip-flop **S–R1** switches over. The logical 0 appears immediately at the output of the final gate AND that has just been active, the signal **P**—a pause—appears, and its count begins by the unit of the delay for turning on. After the pause ends, the flip-flop **S–R2** is set in the new state, the logical 1 appears at the output of the other final gate AND, and the signal **P** is removed. The switching-over comes to an end. With an individual electric drive of the rolls, the rotation speed of the bottom roll is set somewhat higher than the rotation speed of the top roll, in order to obtain an upward bending of the ingot to reduce the impact of the outgoing ingot on the roller table. In addition, the ability to set different rotation speeds can compensate for different roll diameters. However, an increase in the peripheral speed of the bottom roll leads to its overload relative to the load of the top roll, and this overload should not exceed the specified value. Thus, the motor control system of the bottom roll must have the circuits for limiting the speed difference as a function of the difference in the loads of the rolls. Figure 1.7 shows a diagram of setting the rotation speeds. The specified overspeed is assumed to be 2%. If the bottom roll motor current exceeds the top roll motor current by U_0 (assumed $U_0 = 0.4$ pu), the overspeed is reduced.

When modeling the system under consideration, the assignment of the dependence of the difference in rolling torques on the difference in the rotation speeds of the rolls is of great importance. This dependence depends on many factors. On the basis of the data given in Ref. [6], it is assumed in the model that the torque difference of 40% of the total rolling torque appears at a speed difference of 5%, that is, $T_{top} = (0.5 + 4 \times \Delta\omega^*) T_r$, $T_{bot} = (0.5 - 4 \times \Delta\omega^*) T_r$, $\Delta\omega^* = 2(\omega_{top} - \omega_{bot})/(\omega_{top} + \omega_{bot})$, where T_r is the full rolling torque.

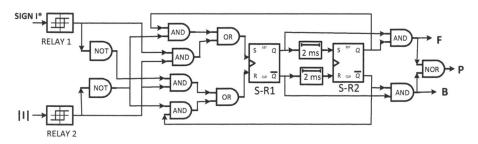

FIGURE 1.6 Diagram of the logical switching unit.

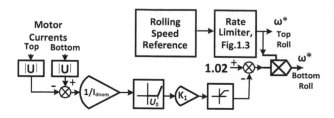

FIGURE 1.7   Diagram of setting the rotation speeds.

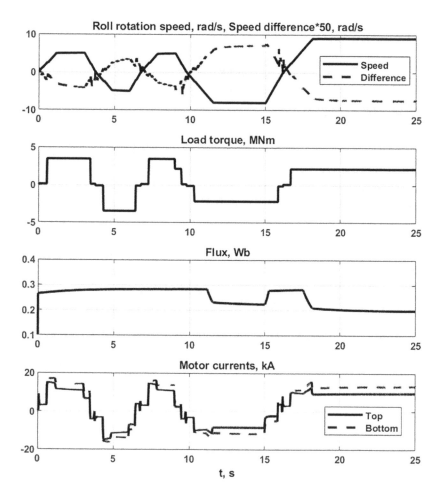

FIGURE 1.8   Processes in the model **Blum2**.

In the **Blum2** model, the electric drive itself is modeled, regardless of the rolling process, the elastic properties of the joints are not taken into account. Start-up and reverse modes are considered when operating at nominal and higher speeds. Figure 1.8 shows some of the simulation results.

Rolling torque is 2.16×1.6 MN-m up to rated speed and decreases to 2.16 MN-m when running with reduced flux. The speed reference changes with a rate of 4.5 rad/ s^2 when accelerating and 7 rad/s^2 when decelerating. Weakening of the flux can be seen when operating at speeds above the nominal. In the upper part of the figure, in addition to the rotation speed of the top roll, the difference in the rotation speeds of the rolls increased by 50 times is shown. It can be seen that at the rotation speed of 9 rad/s, this difference is 0.15 rad/s; 1.7%; this difference is somewhat less than the set value of 2%, since the current difference is 13,200 A − 9500 A = 3700 A or 48%, that is, greater than U_0, which leads to a decrease in the speed difference.

In the **Blum2a** model, the rotation speeds and rolling torques are determined by the assumed rolling program, as in the **Blum1** model. Subsystems for measuring the rms values of motor currents have been added to the schemes of electric drive models. Figure 1.9 shows the bloom rolling cycle. One can observe the change in the motor flux of the bottom, more loaded motor, and its armature current. The rms value of the current during the cycle (40 s), excluding the pause between the ingots, was 9041 A for the top motor and 10,110 A for the bottom one. Thus, the motors are overloaded. Of course, in reality, it is necessary to compare the results of calculations of the rolling torques with the actual values, since they can differ appreciably; at this, one must keep in mind that considered rolling program was used in reality at one of the metallurgical plants [3].

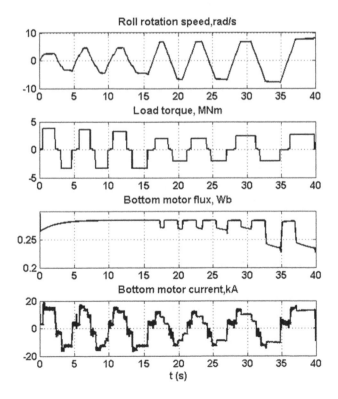

FIGURE 1.9   Processes in the model **Blum2a**.

The **Blum2b** model is the **Blum2** model that takes into account elastic joints in accordance with relations (1.12), (1.14), (1.15), (1.16). The roll speeds are used to distribute the load torques between the motors (instead of the motor speeds). Figure 1.10 shows the plots of the torques transmitted from the motors to the rolls. It can be seen that the elastic oscillations are noticeably larger for the top motor which qualitatively coincides with the results given in Ref. [4].

Starting from the last decades of the last century, almost exclusively AC motors have been used to drive blooming and slabbing rolls [7]. At first, cycloconverters were used to power them, but lately, voltage inverters with synchronous motors have been used for this purpose. Because of the limitations of cycloconverters, such as low power factor, presence of low-frequency harmonics, and less maximum output frequency, advances in IGBT and GTO technology cleared the way for their application to the steel mill drives with PWM two-level and later three-level neutral clamped inverters (VSI) with high switching frequency since the 1990s.

Such systems are modeled in several subsequent models. For simulated blooming, salient–pole synchronous motors are used, having a power of 8 MVA at a voltage of 3150 V. The rotation frequency is 60/100 rpm. Since the motor has ten pole pairs, the rated frequency is 10 Hz. The moment of inertia is 75,000 kg-m^2. Three-level VSI is used to supply motor. DC link voltage is 2×3500 V and is common for both drives. These voltages are produced with the active front rectifier that also is based on the three-level inverter.

Field orientation control principle is used for control of the inverter which supplies motor. Modulus $\Psi_{s\ mod}$ and position θ_e of the stator flux linkage vector are calculated as [8]

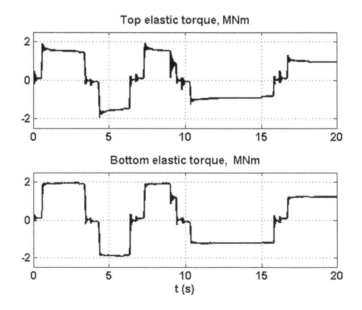

FIGURE 1.10 Elastic torques.

$$\Psi_{s\alpha} = \int \left(U_{s\alpha} + \lambda U_{s\beta} - \lambda \omega_s \Psi_{s\alpha} \right) dt \tag{1.17}$$

$$\Psi_{s\beta} = \int \left(U_{s\beta} - \lambda U_{s\alpha} - \lambda \omega_s \Psi_{s\beta} \right) dt \tag{1.18}$$

$$U_{s\alpha} = V_{s\alpha} - R_s I_{s\alpha}, \ U_{s\beta} = V_{s\beta} - R_s I_{s\beta} \tag{1.19}$$

$$\Psi_{s\,mod} = \sqrt{\Psi_{s\alpha}^2 + \Psi_{s\beta}^2}, \ \theta_e = \text{atan}(\Psi_{s\beta}/\Psi_{s\alpha}), \tag{1.20}$$

where $U_{s\alpha}, U_{s\beta}$ and $I_{s\alpha}, I_{s\beta}$ are the stator voltages and currents.

The vector modulus is used in the voltage regulator of the motor excitation as a feedback signal, and the vector position is used to convert the I_T, I_M components of the stator current into a three-phase signal, which is a reference for the stator current regulator. The setting of the first component I_T^* is determined by the output of the rotational speed controller which is proportional to the wanted motor torque T_m, and the second is equal to zero. The motor flux reference Ψ_s^* is 1 pu as long as the speed reference is less than the nominal speed and decreases inversely with the speed reference increase, so that $I_T^* = T_m/\Psi_s^*$. The maximum value of the torque T_{mlim} is also reduced when the rotation speed is higher than the nominal one. The block diagram of the control system is shown in Figure 1.11 where SC, CC, and FC are speed, motor current and flux controllers, respectively.

In the front active rectifier control system, the current components I_d and I_q of the secondary windings of the transformer are regulated; at this, I_d is directed parallel and I_q is directed perpendicular to the spatial voltage vector of these windings. In this case, the reference of the first value is determined by the output of the voltage regulator at the output of the active rectifier, and the second is equal to zero or, if necessary, can be determined by the output of the reactive power regulator.

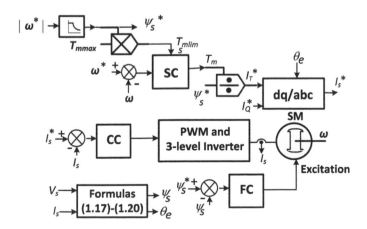

FIGURE 1.11 Block diagram of the control system of the drive with the synchronous motor.

The control system for the speed ratio of the motors of the top and bottom rolls is the same as for DC motors, but instead of currents, the calculated motor torques are used to limit the speed difference.

In the **Blum3** model, the rolling program is not simulated. The speed reference is 6 rad/s, then −8 rad/s, and then 9 rad/s; it varies at rates of 4.5 rad/s² when accelerating and 7 rad/s² when decelerating. The ingot enters the rolls at a speed of 2.5 rad/s, and leaves at a speed of 2 rad/s. The total rolling torque is assumed to be 1.5 × 2.54 MN-m at $t < 6$ s and equal to 2.54 MN-m at $t > 6$ s. The bottom roll speed setting is 2% higher than the top roll speed setting.

The process in the electric drive of the bottom roll with the reactive power controller in the active rectifier is shown in Figure 1.12. It can be seen that, in order to obtain a rotation speed of more than 60 rpm, it is necessary to reduce the stator flux linkage. The active power consumed by both motors from the network can reach 25 MW, and the reactive power is close to zero. The content of higher harmonics in the network current is less than 5%.

In the **Blum3a** model, the rotational speeds and rolling torques are determined by the accepted rolling program, as in the **Blum2a** model.

Figure 1.13 shows the rolling cycle for the electric drive of the bottom roll. It can be seen that its torque reaches 2.5 MN-m in the first passes which is allowed by this motor;

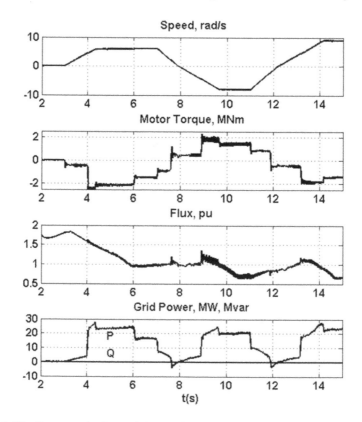

FIGURE 1.12   Processes in the model Blum 3.

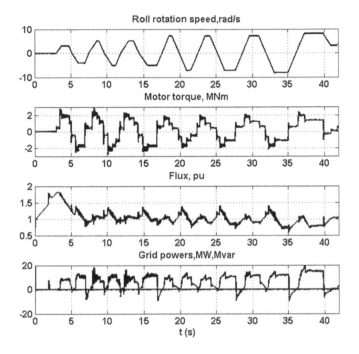

FIGURE 1.13 Processes in the model Blum 3a.

the maximal power consumed from the network reaches 23 MW, and the reactive power is close to zero.

Electric drive of pressure screws (or screw-downs) that set the distance between the rolls, that is, determine the thickness of the metal after rolling, has a significant impact on the productivity of blooming. It consists of two vertical motors, each of which rotates a nut through a gear wheel with a gear ratio i, which causes the screw to move. If d is the average diameter of the screw and α is the helix angle, then, with the motor rotational speed n (rpm), the speed of the screw is $V = \dfrac{n}{60i} \times \pi \times d \times \alpha$. If we take $n = 750$ rpm, $i = 6$, $d = 440$ mm, $\alpha = 2°10'$, then $V = 108.8$ mm/s. Thus, when the motor makes 1 turn, the screws will move 8.7 mm.

The induction motor 570 kW, 690 V, $Z_p = 4$ is accepted. The inertia constant of the motor with attached parts is assumed to be 0.4 s. Rated motor torque is 7.26 kN-m. Moment of resistance when lowering the roll is accepted to be 0.6 pu, and when lifting 0.4 pu. In the model **DTC_pos**, for motor rotational speed control, the DTC version is used, which was developed by the author in Ref. [5] and has some differences from the standard model in the toolbox *Electrical/Specialized Power Systems*. To-day, DTC systems receive a wide application, therefore it is supposed that the reader is familiar with the principles of operation of such systems, and they are not further described in detail.

The model runs in pu that simplifies its use when the drive parameters change; the circuit in the dynamic braking circuit is controlled by the simpler relay circuit.

For torque control, two-position relay instead of three-position scheme is employed. Depending on the value and sign of the rotational speed ω, three tables are used for choosing the next inverter state. When $|\omega| > \omega_0$ where ω_0 is a small value, for example, $\omega_0 = 0.05$ pu, one of two tables is selected, depending on sign ω, in which zero states of the inverter are chosen, when it is necessary to decrease the torque, which leads to a reduction in the torque pulsation. When $|\omega| < \omega_0$ the third table is selected, in which only active states are used. For flux calculation, the formulas (1.17–1.20) are used.

An active rectifier is used to power the DC link, the circuit, and the control system of which are described in the previously considered models. The position controller has a non-linear characteristic, the equations of which, bearing in mind some simplifying assumptions, are derived as follows. If ΔQ_d is the deviation of the position of the motor shaft from the given one, and ω is its angular rotational speed, then, to obtain a constant deceleration, it should be $\omega^2 = 2a\Delta Q_d$, where a is the deceleration value that equal to T_{din}/J, T_{din} is a taken braking dynamic torque, J is the total moment of inertia. Thus, taking into account that the rotational speed is given in pu, and the path is given in rad, the required dependence of the rotation speed on the remaining error during braking has the form

$$\omega^* = \sqrt{\frac{2\Delta Q_d T_{din}}{J\omega_b^2}} = \sqrt{\frac{2\Delta Q_d T_{din}^* P_b}{J\omega_b^3}} = \sqrt{\frac{\Delta Q_d T_{din}^*}{H\omega_b}} \tag{1.21}$$

And taking $T_{din}^* = 1$, $H = 0.4\ s$, $\omega_b = \dfrac{314}{Z_p} = \dfrac{314}{4}$, we receive $\omega^* = \sqrt{\dfrac{\Delta Q_d}{31.4}}$.

At some error ΔQ_0, the non-linear characteristic should turn into a linear one with a slope K_l. To maintain continuity, the relation $\sqrt{\dfrac{\Delta Q_0}{31.4}} = K_l \Delta Q_0$ must hold, from which $K_l = 1/\sqrt{31.4\Delta Q_0}$.

It is assumed in the **DTC_pos** model that the position sensor is connected directly to the motor shaft, so that the roll position S is converted to the shaft position Q as $Q = S \times 2\pi/8.7$. It is taken $\Delta Q_0 = 0.1$ rad.

The process of respat the following positions: 0, 20 mm, 160 mm, 0 is shown in Figure 1.14. It can be seen that the transient processes occur without overshoots. When the next transient ends, the motor remains under load, which, on the one hand, increases motor heating, but, on the other hand, prevents an uncontrolled change in position caused, for example, by an attempt of self-unfastening the screws during rolling.

Next, the joint operation of the pressure screws with the main drive is simulated. To speed up the simulation, a simplified main drive model from the **Blum1** model is used. Of course, other complete models developed above can also be used if necessary. When moving, there may be some discrepancy in the position of both screws, caused by the difference in characteristics and forces of resistance, which can be eliminated by introducing a displacement difference controller, but in this case, it does not matter, since both screws are at rest in the same position during rolling. Therefore, only one screw is modeled.

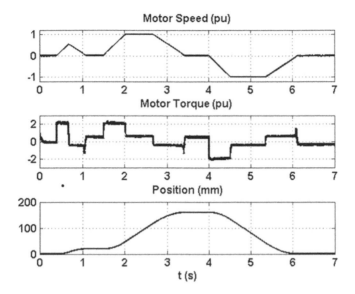

FIGURE 1.14 Respat the screw-down positions.

The **Blum4** model has some changes compared to the original model. Thus, the setting of a new screw position occurs immediately after the ingot leaves the rolls (the **Hout** subsystem). After the end of the last, 13th pass, the initial setting of the position is 840 mm. After the main motor is reversed, its speed does not exceed 2 rad/s until the pressure screws are in the required position (*Stop* signal is not equal to 0).

Figure 1.15 shows the process of reversing the main drive for the 12th pass. At point *a*, the speed of its rotation begins to decrease, and at this moment the screw position reference for the next pass is formed. At point *b*, the ingot leaves the stand, the load torque decreases to the idle one, and the pressure screws start working out a new position. At point *c*, the main motor begins to rotate in the opposite direction without load, and at point *d* it reaches a speed of 2 rad/s; no further increase in speed occurs, since the pressure screws have not yet taken the required position. This occurs only at point *e*, after which the rotational speed of the rolling motor begins to increase, at point *f* the ingot enters in rolls, and at point *g* the motor reaches the given speed of 8 rad/s in the direction opposite to the initial one. Figure 1.16 shows the ingot rolling cycle.

1.1.2 PLATE ROLLING MILL ELECTRIC DRIVES

In modern rolling production, slabbing mills are rarely used, since slabs for sheet rolling mills are produced in continuous casting machines. At the same time, old plate mills are being modernized and new ones are being built due to the increasing demand for wide hot-rolled sheets for nuclear power engineering, for the construction of large tankers, for the production of large-diameter pipes, and so on. New plate mills are characterized by an increased length of the roll barrel—up to 5–5.5 m, the high stiffness of the rolling stand—up to 10–12 MN/mm, which allows a rolling force of up to 100 MN.

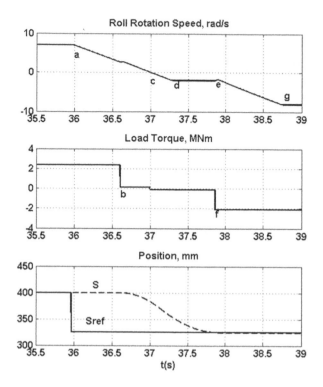

FIGURE 1.15   Process of reversing the blooming main drive.

The stands with horizontal rolls are four-roll (Figure 1.5b), and with an individual electric drive of the rolls, the motor power reaches 10–12 MW. Plates at the output have a thickness of 5–50 mm or more, with a width of 4.8–5.3 m and a length of up to 60 m. Since the width of the input ingot is usually much less than the required width of the plate, several passes are made in the rolling stand in the crosswise direction, in order to increase the width of the plate to be rolled.

Mills can have one, two, or three stands. In the two-stand mill, one of the stands can be with vertical rolls (edge stand) or both stands with horizontal rolls, roughing and finishing; in the first one, a plate of a wanted width is formed, and 70%–75% of the total reduction of the plate is performed, and in the second stand, the remaining 20%–25% is performed, with control of the output thickness. The reductions between the stands are distributed in such a way that the total rolling time in the finishing stand would be somewhat less than the rolling time in the roughing stand, so that at the end of the rolling of the next ingot in the roughing stand, the finishing stand would be free and ready to continue rolling without delay.

In a three-stand mill, one of the stands is always with vertical rolls; it is installed in front of the roughing stand and is designed for descaling, for forming edges, and for adjusting the width. After several passes in the roughing stand, the plate is transferred to the edge stand for some width change and then returns to the roughing stand to continue rolling. The plate in each stand is rolled independently of the other stands, therefore, when modeling, it is sufficient to simulate one stand, which is currently of interest.

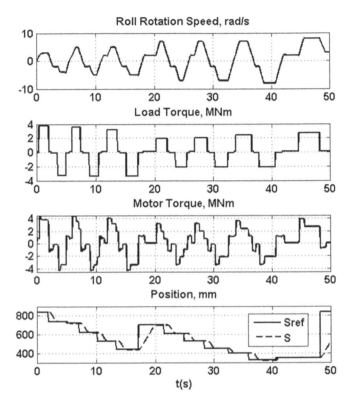

FIGURE 1.16 Ingot rolling cycle.

Next, a stand is modeled with the following parameters: roll diameters: 1200/2300 mm, length 5000 mm, stiffness modulus 10 MN/mm, each roll is driven by a synchronous motor 12,000 kW, 60/115 rpm, nominal torque 1.91 MN-m, short-term permissible overload 2.5. The moment of inertia of the motor with the attached parts can be assumed to be 200,000 kg-m². Acceleration during speeding-up with metal is taken equal to 1.5 rad/s². Assuming the loss torque equal to 5% of the nominal, the maximal rolling torque at a speed below the nominal can be estimated as $2 \times (2.45 \times 1.91 - 1.5 \times 0.2) \approx 8.7$ MN-m. As an example, rolling of the plate 10×4800 mm from the ingot $250 \times 2400 \times 4700$ mm, alloy steel (USA: 3415, DIN: 1.5732) is modeled.

For calculation, the program *Plt1.m* is used which is a slightly modified *Blm1.m* program; in particular, the yield strength for the taken steel is calculated by the formula [1]

$$\sigma_s = \frac{2300\varepsilon^{0.252}u^{0.143}}{e^{0.0029T}} \ \text{N/mm}^2 \tag{1.22}$$

The initial temperature is assumed to be 1100°C and decreases after each pass by [3]

$$\Delta T = 0.0021\left(\frac{T+273}{100}\right)^4 \frac{\tau}{h_1} \tag{1.23}$$

where h_1 is the thickness after the pass, τ is the cooling time equal to the rolling time plus the pause time. A program for the calculation of the rolling parameters in the 8th pass is given below.

Program *Plt1*

```
hin=70;%input thickness,mm
hout=50;%output thickness,mm
R=600;%roll radius,mm
T=1063;%initial temperature,gtad C
f=0.8*(1.05-0.0005*T);%friction
dH=hin-hout;
hm=(hin+hout)/2;
l=sqrt(R*dH)%arc of contact
W0=4762;%initial width,mm
dW=1.15*dH/2/hin*(1-dH/2/f);%width changing,mm
if hin/W0<1
    W1=W0+dW;%output width,mm
else
    W1=W0+dW*(1+(hout+0.5*dH-f*l)/(2*R));
end;
Wi=W0+dW/2;
r=dH/hin;
d=l/hm;
Vp=3.5;%rolling speed, m/s
wr=1000*Vp/R
u=1000*Vp*r/l;%deformation speed
s1=2300*r^0.252*u^0.143/exp(0.0029*T)%N/mm2
S=1.15*s1
n1=1+f*d/2
n2=d^(-0.3)
Pputs=S*n1*n2%average unit rolling force,N/mm2
N=wr*30/pi%rotational speed, rpm
Pts=Pputs*Wi*l%rolling force, N
Mts=2*Pts*l/1000*0.5%rolling torque, Nm
Powerts=Mts*wr/1000%rolling power, kW
tau=12;%cooling time, s
dT=0.0021*(0.01*T+2.73)^4*tau/hout
L0=2400;%initial ingot length, mm
L1=L0*250*4700/Wout/hout %sheet length after rolling
alpha=sqrt(dH/R)*57.3%angle of bite, degree
```

To obtain a plate of the required width, after rolling in a stand, it is additionally rolled in a vertical stand (if any) or processed on the side trimming shears.

In the **Plate1** and **Plate2** models, the mill is simulated in a simplified way, as in the **Blum1** and **Blum 4** models; these mentioned models have been modified as necessary to accommodate changes in the parameters of the mill and rolled metal. The most significant change is the introduction of the **Temperature** subsystem, in which the calculation of the rolled ingot temperature in the given pass is performed using formula (1.23); the value of τ is fixed in the **Sample and Hold 2** block, and the calculated temperature in **Sample and Hold 3**. The motor acceleration is taken equal to 1.5 rad/s² in the first six passes with an increase to 3 rad/s² in the subsequent ones; deceleration is 3 rad/s². This model can be used to check the correctness of the calculation of the rolling parameters and the functioning of the system as a whole. Figure 1.17 shows the process of rolling one ingot. Some discrepancies between the values of the rolling torques and the given ones in Table 1.4 are due to the fact that in the latter, the values of τ were taken to some extent arbitrarily when calculating the temperature.

Plate2 model contains the model of the electric drive of the pressure screws, borrowed from **Blum4** model, but the speed of the screws is taken half as much, for example, the gear ratio of the reduction gear is increased from 6 to 12. The thickness of the plate at the exit of the stand is determined by the relation

$$h_1 = S_0 + \frac{P}{K} \tag{1.24}$$

where S_0 is a roll gap when the plate is absent, P is a rolling force, K is the stiffness of the rolling stand. Thus, to obtain the required thickness, the position of the pressure screws must be reduced by the amount of P/K. As for the value of P, a theoretically calculated quantity can be taken (in practice, previously measured and recorded values). The results of the calculations given in Table 1.4, subsystem **Hout**, are used for this purpose in the **Plate 2** model. Figure 1.18 shows the same process,

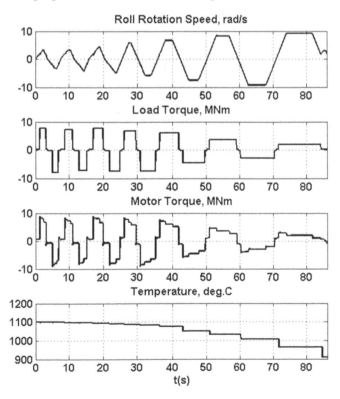

FIGURE 1.17 Plate mill rolling cycle.

TABLE 1.4

Rolling Program of the Ingot 250 × 2400 × 4700 mm in the plate 10 × 4800 mm in 13 Passes

Pass	h_0 (mm)	W_0 (mm)	L_0 (mm)	h_1 (mm)	W_1 (mm)	L_1 (mm)	$\alpha°$	V (m/s)	M_{ts} (MNm)
1	250	4700	2400	220	4707	2723	13	2.0	7.7
2	220	4707	2723	190	4714	3148	13	2.0	8.0
3	190	4714	3148	163	4722	3663	12	2.0	7.3
4	163	4722	3663	138	4730	4319	12	3.0	7.8
5	138	4730	4319	113	4740	5263	12	3.0	7.7
6	113	4740	5263	90	4751	6596	11	3.5	7.9
7	90	4751	6596	70	4762	8460	10	3.5	7.2
8	70	4762	8460	50	4775	11,811	10	3.5	8.0
9	50	4775	11,811	35	4788	16,826	9	4.0	6.6
10	35	4788	16,826	25	4799	23,506	7	4.5	4.8
11	25	4799	23,506	18	4808	32,584	6	5.0	3.8
12	18	4808	32,584	13	4816	45,043	5	5.5	3.5
13	13	4816	45,043	10	4821	58,491	4	5.5	2.4

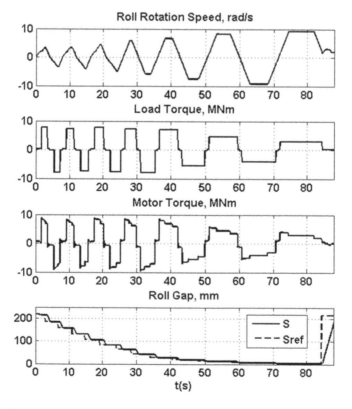

FIGURE 1.18 Plate mill rolling cycle including pressure screws function.

as in Figure 1.17, obtained in the **Plate 2** model. An examination of the fixed process shows that, with the taken parameters, the movement of the pressure screws, which begins after the plate leaves the stand, ends, in most passes, before a new bite of the plate occurs, when the rolls rotate in the opposite direction with the speed corresponding to 2 m/s. Thus, taking into account the operation of the pressure screws does not affect the processes in the model of the electric drive of the rolls; at the same time, modeling in the **Plate 1** model is several times faster than in the **Plate 2** model.

In the **Plate 1a** model, circuits have been added to the **Rolling_Parameters** subsystem to study changes in the plate thickness. When the output thickness h_1 changes, the rolling force P changes too, which in turn affects h_1, so that a so-called algebraic loop is formed; in order to break this loop, a block with a first-order transfer function with small time constants is used. It is assumed that the unloaded roll gap is determined by the sum $S_0 = S_{scr} + S_h$, where the first value is determined by the position of the pressure screws, and the second by the displacement of the hydraulic system for the bottom roll installation. At that, the first value is determined by (1.24) for estimated values of the rolling force and stand stiffness, as in the **Plate 2** model, and does not change in the running pass.

If the actual value of the rolling force is not measured, then, as a result of the simulation (the **Switch** tumbler is in the left position), assuming the exact value of K is known, alternations in the plate thickness after the last pass is shown in Figure 1.19a.

If the rolling force is measured, then it is possible to create indirect thickness control using relation (1.24), acting on the hydraulic system of the bottom roll installation.

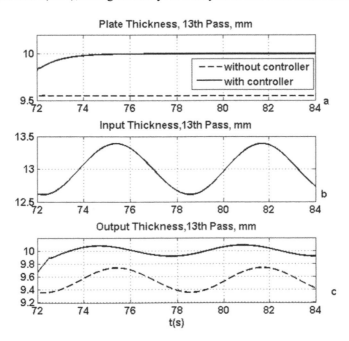

FIGURE 1.19 Thickness change after the last pass. (a) Without and with thickness control; (b) periodic thickness deviations; (c) Without and with thickness control under periodic thickness deviations.

The model of such a controller is in the **Controller** subsystem, whose structure is shown in Figure. 1.20. The hydraulic drive is modeled in a simplified way, as a serial connection of the unit with the first order transfer function and the roll speed displacement limiter. The time constant of the hydraulics is taken equal to 30 ms, and the maximum speed of the roll movement is 1 mm/s. The change in the plate thickness after the last pass with the operation of the indirect thickness controller is also shown in Figure 19a. It can be seen that the wanted thickness value of 10 mm is provided.

The change in the thickness of the plate after the last pass is shown in Figure 1.19c, when the plate at the entrance to the stand has periodic thickness deviations from the nominal value of 13 mm, shown in Figure 1.19b. It can be seen that the deviations of the thickness from the wanted one are significantly reduced.

It should be said that the control of the thickness of the rolled plate on a plate mill is a difficult task due to the presence of transverse thickness variations and the inability to make measurement in proper time and in a sufficient number of points. Some approaches to solving this problem are described in Refs. [9,10], but they are beyond the scope of this book.

The models of the synchronous motors with their supply and control systems are included in the following models. The electrical drive of one roll, where the motor is supplied by three-level VSIs, is simulated in the model **Plate3**. To decrease simulation time, DC link voltage is taken constant 2×3500 V. In the currently in-operation plate mill, with motors 12,000 kW, three inverters are connected in parallel [11], as shown in Figure 1.21.

To reduce the higher harmonic content in the motor current, the triangular carrier signals for PWM for each inverter are shifted by 1/3 of a period relative to each other. Since the standard PWM model for the three-level inverter does not provide for a phase

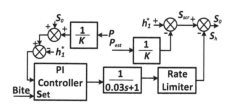

FIGURE 1.20   Thickness control system.

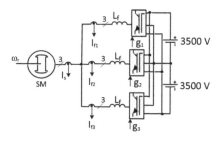

FIGURE 1.21   12,000 kW motor power circuits.

shift of the carrier signals, a PWM block has been developed for this model, included in the **Vector control/Current regulator** subsystem. In other respects, the same control system is used as in the **Blum3** model, with a corresponding change in some parameters. In the considered model, the motor accelerates to a nominal speed of 6.28 rad/s, the stator current frequency is 10 Hz, with a torque of 0.5 nominal, and at $t = 8$ s, the load increases to the nominal. The inductances of the reactors are taken equal to 0.0625 mH, and the frequency of the carrier signal is 500 Hz. In this case, the stator current is almost purely sinusoidal: the Total harmonic distortion (THD) value is about 1%. If the simulation is performed in the absence of a shift between the carrier signals ($g^2 = g^3 = g^1$), then we get THD $\approx 4\%$. Thus, the phase shift of the carrier signals is quite effective.

The circuits for elastic torque simulation are added in the **Plate 3a** model; these circuits are taken from the **Blum2b** model. Data are given for the rolls of the 5000 plate mill in Ref. [12]: the work roll: diameter 1.2 m, weight 63,000 kg; back-up roll: diameter 2.3 m, weight 226,400 kg. Thus, the moment of inertia of the load reduced to the motor shaft is $J_2 = 0.125 \times (63,000 + 226,400) \times 1.2^2 \approx 50,000$ kg-m^2. Since earlier the total moment of inertia was taken equal to 200,000 kg-m^2, the inertial constant of the motor is reduced by 25% when elastic links are taken into account. Because the motor maximal torque can reach 4.3 MN-m, taking double margin, we receive from (1.15) $D > \sqrt[3]{\dfrac{16 \times 2 \times 4.3 \times 10^6}{\pi \times 80 \times 10^6}} = 0.82$ m. It is accepted $D = 0.9$ m. In the plate mill 5000, the bottom motor is closer to the stand; let's take for the top roll $l = 16$ m and for the bottom one $l = 8$ m, so that

$$C_{top} = \frac{\pi \times 80 \times 0.9^4 \times 10^9}{32 \times 16} = 322 \text{ MN-m/rad and}$$

$$C_{bot} = \frac{\pi \times 80 \times 0.9^4 \times 10^9}{32 \times 8} = 643 \text{ MN-m/rad}$$

The model also provides the opportunity to simulate backlash in the spindle. The model allows performing a series of experiments to determine the influence of mechanical transmission properties on the characteristics of the electric drive. As an example, Figure 1.22 shows the process when the motor accelerates to a rotational speed of 2.5 rad/s without a load, after receiving the load it accelerates to the rated speed, goes into braking mode, unloads at a speed of 1.5 rad/s, then rotates at a speed of 2.5 rad/s in the opposite direction, waiting for the plate new bite. The backlash value is taken equal to 1°. Noticeable fluctuations in the torque are visible when the plate is captured or leaves the rolls and at the moment of revers.

Various circuits can be used to generate DC voltage for powering inverters. In the 5000 plate mill drive, each of the three inverters of each motor is powered by a separate active rectifier, which forms a separate DC link. Each active rectifier is powered by a separate transformer, the primary windings of these transformers are connected in parallel to the mains voltage, and the windings of three transformers of each motor provide phase shifts of 0°, 20°, −20°, in order to reduce the content of higher harmonics in the mains current [11], Figure 1.23.

Such a system is simulated in the **Plate32** model. Inductances L_g are taken equal to 0.6 mH. Transformers Tr1–Tr3 have a power of 8 MVA each at a voltage

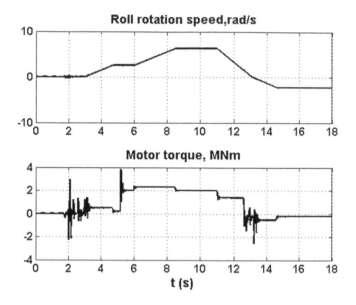

FIGURE 1.22 Torque fluctuations when elastic coupling is taken into account.

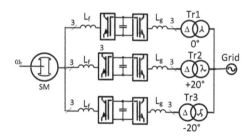

FIGURE 1.23 DC voltage generation in the system with three transformers and three active rectifiers.

of 35/3.2 kV. The voltage in the DC link is 2×4200 V. The frequency of the carrier signal is 600 Hz. The process of simulation in the model is slow, so Figure 1.24 shows only a fragment of the process, when the motor speeds up with an acceleration of 3 rad/s² to a rotation speed of 2π rad/s with a load of 0.5 nominal, at $t = 5.1$ s it reaches a steady speed, and at $t = 5.5$ s receives the rated load. It can be seen that the grid current at full load is close to sinusoidal (THD = 3.2%), while the currents in transformers have THD $\approx$ 6%.

One transformer 20 MVA feeds all three active rectifiers in the **Plate33**.model, Figure 1.25. The same process, as in Figure 1.24, is shown in Figure 1.26, the carrier signal frequency is 1200 Hz. It can be seen that the content of higher harmonics in the network current has increased noticeably (THD has increased to 8%), so in practice, additional measures may be required to improve the shape of the network current, which can substantially level the simplification of the main circuits associated with a decrease in the number of transformers.

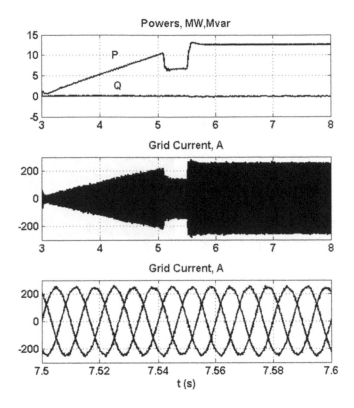

FIGURE 1.24   Processes in the **Plate 32** model.

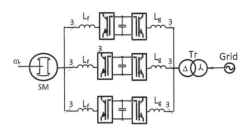

FIGURE 1.25   DC voltage generation in the system with one transformer and three active rectifiers.

Possible simplifications of the system are associated with the introduction of common DC buses with voltage U_d, Figure 1.27. In the **Plate34** model, the voltage regulator U_d is located in the **Rectifier1** subsystem and generates current reference signals for the active rectifier current regulators in the **Rectifier1-Rectifier3** subsystems. When simulating the same process, as in the previous two models, THD = 4% is obtained in the mains current with the carrier frequency of 600 Hz.

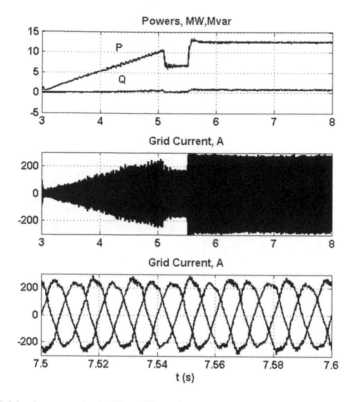

FIGURE 1.26   Processes in the **Plate 33** model.

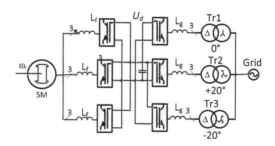

FIGURE 1.27   Common DC buses for the inverters of the plate mill drives.

1.2   ELECTRICAL DRIVES OF THE CONTINUOUS HOT STRIP MILLS

1.2.1   ROUGHING MILL GROUP DRIVES

Continuous hot rolling mills consist of roughing and finishing groups. The distance between the stands of the roughing group is large enough (for example, 12 m), so that the sheet passes through the stands independently. In this case, there are no special specific issues related to technology. To drive the rolls, both adjustable (direct and alternating current) motors and unregulated synchronous motors are used.

However, recently, the situation has changed: the last two or three roughing stands are combined into a continuous group, which makes it possible to increase the rolling temperature and reduce the size of the shop. Here we consider the case when a continuous roughing group consists of three stands with adjustable speeds.

All three stands are 4-roll with a working roll diameter of 1200 mm. Sheet width $b = 1600$ mm. The rolling speed at the exit of the last stand is 5 m/s. As a typical rolling schedule, the following is accepted: 124–79 mm, 79–54 mm, 54–40 mm. In this case, the reduction value should not exceed the values allowed by the bite conditions (1.1) with $\alpha = 16°–17°$, $\Delta h_{max} = 600 \times 0.28^2 = 47$ mm. The stands are driven by the same synchronous motors as for plate mills: 12,000 kW, 60 rpm, the number of the pole pairs $Z_p = 10$, the nominal frequency is 10 Hz.

There are various methods for calculating specific rolling forces, and it is impossible to say, which one is more consistent with the actual values. The choice of the applied technique is determined by preferences, application experience, convenience, and so on. In the previous section, the technique described in Ref. [2] was applied. To calculate the rolling forces in the continuous hot strip mills, the Sims method, described, for example, in Ref. [13], has found application. At this, $p_a = 1.15\sigma_s Q_p$, where

$$Q_p = a\left[\frac{\pi}{2}\mathrm{atan}(1/a) - \frac{\pi}{4a} - a_2 \times \log_e \frac{Y}{h_1} + 0.5a_2a_3\right] \qquad (1.25)$$

$$a = \sqrt{\frac{1-\varepsilon}{\varepsilon}}, \; \varepsilon = \frac{\Delta h}{h_0}, \; \Delta h = h_0 - h_1, \; a_2 = \sqrt{\frac{R}{h_1}}, \; a_3 = -\log_e(1-\varepsilon).$$

Here $Y = h_1 + Rc^2$, $c = \dfrac{\tan(a_1 + b)}{a_2}$, $a_1 = -\dfrac{\pi}{8}\dfrac{a_3}{a_2}$, $b = 0.5 \times \mathrm{atan}\left(\dfrac{1}{a}\right)$.

When calculating the rolling torque according to (1.10), the arm of force is calculated more accurately, namely [2]:

$$\psi = 0.5\left(1 - \varepsilon\frac{1+m}{2+m}\right), \; m = \frac{2fl}{h_0 + h_1}. \qquad (1.26)$$

Friction coefficient f is computed by (1.6) and $l = \sqrt{R\Delta h}$. Calculation of σ_s is carried out by (1.2). The program $Sims1$ computes rolling force and torque using these relations.

Program $Sims1$

```
hin=124;%stand input thickness
hout=79;%stand output thickness
hend=40;%roughing group thickness
Vend=5;%roughing group speed,m/s
R=600;%roll radius
T=1280;%temperature, degree C
Wi=1600%sheet width
f=0.8*(1.05-0.0005*T);
```

```
dH=hin-hout;
hm=(hin+hout)/2;
r=dH/hin;
l=sqrt(R*dH);
d=l/hm;
m=f*d;
Vp=Vend*hend/hout;
wr=1000*Vp/R;
u=1000*Vp*r/l;
a=sqrt((1-r)/r);
a2=sqrt(R/hout);
a3=-log(1-r);
a1=-(pi/8)*a3/a2;
b=0.5*atan(1/a);
c=tan(a1+b)/a2;
Y=hout+R*c*c;
Qp=a*(pi/2*atan(1/a)-pi/4/a-a2*log(Y/hout)+0.5*a2*a3)
s1=1330*r^0.252*u^0.143/exp(0.0025*T)%N/mm2
S=s1;
Ppus=1.15*S*Qp*l;
Ps=Ppus*Wi%rolling force,N
Tr=Ps*l/1000*(1-r*(1+m)/(2+m))%rolling torque, Nm
Powers=Tr*wr/1000%rolling power,kW
N=wr*30/pi%rotational speed, rpm
```

Let us use this program to calculate the power of the roughing stands with the input temperature of 1280°C. For the first stand one receives the roll force $P = 18.4\,MN$, the roll torque $T_r = 2.36$ MN-m, $\sigma_s = 53.7$ N/mm², $N = 40.3$ rpm. The rolls are driven by the abovementioned motor with help of a reduction gear 1.5 (Figure 1.5c), which provides a nominal torque of 2.87 MN-m on the rolls at a nominal rotation speed of 40 rpm.

For the second stand $P = 15.2\,MN$, $T_r = 1.5\,MN$-m, $\sigma_s = 56$ N/mm² $N = 58.9$ rpm. The motor rotates the rolls without gear (Figure 1.5d). For the third stand, under T = 1260°C, $P = 12.4$ MN, $T_r = 0.96$ MN-m, $\sigma_s = 68.8$ N/mm², $N = 79.6$ rpm. The motor operates with flux reduction and provides the torque $1.91 \times 60/79.6 = 1.44$ MN-m.

When the sheet is in several stands, there are interstand tensions, which may have a minus sign (there is a compression), since the sheet is thick enough, which tell on the rolling process as follows:

The rolling force changes; in the formula for p_a, instead of 1.15 σ_s, one can take $1.15\sigma_s - 0.5\,(\sigma_0 + \sigma_1)$, where the last two values are the specific back and forward tensions, moreover, if the previous stand is accelerating, then $\sigma_0 < 0$, and if the subsequent stand accelerates, then $\sigma_1 > 0$, and vice versa.

The speed of the sheet exiting the roll bite changes at the same circumferential speed of the rolls, as the forward slip s changes [13]:

$$s = \left(\frac{R}{h_1}\right)\beta^2 \tag{1.27}$$

$$\beta = \left(0.5 \sqrt{\frac{\Delta h}{R}} - \frac{\Delta h}{4Rf} \right) + \frac{\sigma_1 h_1 - \sigma_0 h_0}{4Rf p_a} = A + B(T_1 - T_0) \qquad (1.28)$$

Here T_0, T_1 are full sheet tensions, $B = 1/(4Rbf p_a)$, b is the sheet width, p_a is the specific rolling force without tensions. Let us estimate the quantities in this relation.

For the first stand, $f = 0.8 \times (1.05 - 0.0005 \times 1280) = 0.33$, $A = 0.5 \sqrt{\dfrac{45}{600}}$

$$-\frac{45}{4 \times 600 \times 0.33} = 0.08, \; B = \frac{1}{4 \times 600 \times 1600 \times 0.33 \times 70.4} = 11.2 \times 10^{-9}.$$

In actual conditions, the maximum tension values do not exceed 15%–20% of the yield strength, so the second term in the formula for β is much less than the first one, and a linear dependence can be taken as

$$s = \frac{R}{h_1} \left(A^2 + 2ABΔT \right) = s_0 + γΔT \qquad (1.29)$$

$ΔT = T_1 - T_0$; in the case under consideration, $s_0 = \dfrac{600}{79} \times 0.0064 = 0.049$, $γ = 13.6 \times 10^{-9}$, and, under specific forward tension 10 N/mm² (back tension for the first stand is 0), $s = 0.066$ Analogous, for the second stand $s_0 = 0.054$, for the third stand $s_0 = 0.052$.

— Direct influence of tensions on the rolling torque T_r may be written as

$$T_r = T_d + R\left(\frac{T_0}{\lambda} - T_1 \right)(1 + s) \qquad (1.30)$$

where T_d is a torque, demanding for sheet deformation under given conditions, λ is the sheet elongation, $\lambda = h_0/h_1$.

Tension change between stands $i + 1$ may be written as

$$\frac{dT_{i,i+1}}{dt} = \frac{EQ_i}{L} \left(V_{0,i+1} - V_{1,i} \right) \qquad (1.31)$$

where E is the modulus of elasticity (Young's modulus), $E = 0.2$ MN/mm², Q_i is the cross-sectional area of the sheet after the i-th stand, L is the distance between the stands, $V_{0,i+1}$ is the speed of the sheet entering the $i + 1$ stand, which can be written as $V_{0,i+1} = \dfrac{V_{1,i+1}}{\lambda}$, $V_{1,i}$ is the speed of the sheet exiting the i-th stand that is equal to $V_{1,i} = \omega_{r,i} R_i \left(1 + s_i(T) \right)$, where $\omega_{r,i}$ is the rotational speed of the rolls of the i-th stand, R_i is their radius, s_i is the forward slip in the i-th stand, depending on the tension.

Rolling in the roughing group should be carried out without tension between the stands; at the same time, there are no direct or indirect meters of this tension. Therefore, the following approach is used: some process parameters are fixed, which change, when interstand tension appears, and the stand drive control system affects the rotation speeds of the stand motors in such a way that these parameters remain constant when there is a possibility of occurrence of interstand tensions. As such a parameter, it is proposed in Ref. [14] to take the ratio of the motor torque to the value of the rolling force, that is, the value $Z = T_r/P$, or the reciprocal of $Y = P/T_r$. The introduction of dividing by P is due to the fact that the rolling torque can change not only due to a change in tension but also due to a change in the rolling force (changes in metal hardness, changes in input thickness).

The roughing group consisting of two stands is considered in the **Mill_rough1** model, the speed of rotation of the first stand motor is controlled. The parameters of the stands and the rolling process correspond to stands 1 and 2 of the three-stand group described above. In this model, the motors with their power systems are modeled in a simplified way, as in the **Plate1** model, this model is intended to confirm the principles of regulation and to determine the main parameters. Let us compute the inertia constants when the motor moment of inertia $J_m = 125{,}000$ kg-m^2 and take the roll moment of inertia equal to $0.15\,J_m$. For the first

$$\text{stand, } H = \frac{1.25 \times 10^5 \left(1 + \dfrac{0.15}{2.25}\right) \times 6.28^2}{2 \times 12 \times 10^6} = 0.22\,\text{s, for the second and the third stands}$$

$$H = \frac{1.25 \times 10^5 \times 1.15 \times 6.28^2}{2 \times 12 \times 10^6} = 0.24\,\text{s}.$$

Calculations of the rolling torques taking into account the sheet tension are performed in the subsystems **Torque_Stand1** and **Torque_Stand2** (subsystem **Load**). These calculations are relevant when the sheet is in the rolls (signals *bite1*, *bite2*). The first signal appears after the end of the acceleration at $t = 2$ s, and the second one later by $12/V_1$ s, where 12 m is the distance between the stands, V_1 is the exit speed from the first stand. The sheet tension T_{1-2} is calculated in the **Tension** subsystem using the above formulas for tensions and forward slips after the sheet enters the second stand. The quotient of dividing two quantities is computed in the **Tension_Control** subsystem, the first quantity is the motor torque of the first stand (in practice, the scheme for its calculation depends on the structure of the electric drive control system), and

the second one is $P' = P\left(1 - \dfrac{T_{1-2}/bh_1}{1.15\sigma_s}\right)$, where P is the rolling force when the tension

is absent. The resulting quotient value is stored in the **Sample&Hold** block at $t = 5$ s, that is, when the sheet is about halfway to the second stand. By command of the load current relay of the second stand, which fixes the sheet entry into the second stand (when the motor torque exceeds 25% of the nominal value, signal T_{ref2}), the difference between the stored and current values of the quotient begins to be integrated: the integrator output is added to the speed reference of the first stand, reducing the difference of these values to zero, and, consequently, reducing the tension of the sheet to zero. The described process is shown in Figures 1.28 and 1.29. In the first case, the motor speed of the second stand is set 2.5% higher than it follows from the condition of the volume constancy, and in the second case, it is 2.5% lower, which causes significant

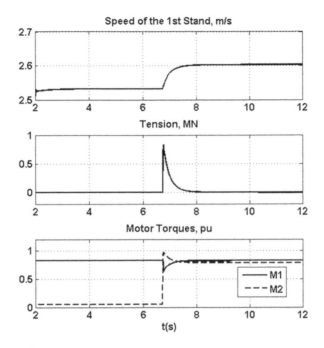

FIGURE 1.28 Change in tension when the sheet enters the second stand and. its speed is higher than required.

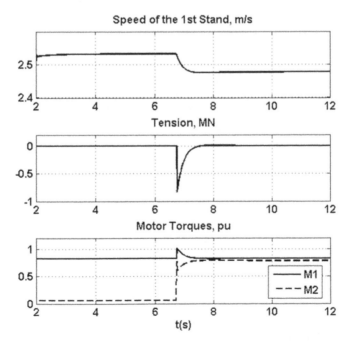

FIGURE 1.29 Change in tension when the sheet enters the second stand and its speed is less than required.

interstand tension when the sheet enters the second stand, but it is parried by the described control system.

A roughing group consisting of the three stands described above is considered in the **Mill_rough2** model. The speeds of rotation of the first and third stand motors are regulated, and the second stand serves as the base one. In this model, the motors, with their power systems are also modeled in a simplified way. Circuits have been added to the electric drive model that takes into account the possible weakening of the motor field: when a rotation speed is more than 1 pu, the motor torque reference decreases, under the same value of the speed controller output, which, with the same load torque, leads to an increase of the controller output, which determines the active component of the motor current. The **Torque_Stand3** subsystem has been added to the **Load** subsystem, in which the rolling torque of the third stand is calculated when the sheet appears in this stand (the *bite3* signal); after the sheet has been captured by the second stand, its speed begins to be integrated, determining the path traveled by the sheet; the *bite3* signal is generated when this path becomes 12 m. Simultaneously, the sheet tension T_{2-3} is calculated in the **Tension** subsystem. The added **Tension-Control3** subsystem has the same structure as the **Tension-Control1** subsystem, but both forward and back tensions are taken into account. The given signal is stored when the sheet is halfway to the third stand. By the command from the load current relay of the third stand, a signal for correction of the rotation speed of the third stand appears. Note that the signs of the correcting signals are opposite for the first and third stands, since they are on opposite sides of the base stand.

The process is shown in Figure 1.30. The motor speed of the second stand is set 2.5% higher than it follows from the volume constancy relation. Thus, when the sheet enters the second stand, tension 1–2 increases, and when the sheet enters the third stand, tension 2–3 decreases (get negative). These deviations are parried by the described control system. It can be seen that, due to the weakening of the field of the stand 3 motor, its current in pu exceeds the torque value.

Mill_rough3 model models the roughing group with the models of the synchronous motors 12 MW, whose power and control systems are taken from the model **Plate3**. Simulation in such a system, even with increased simulation discreteness (10 μs), is slow, so some simplifications are used. The **Mill_rough3** model assumes that the DC link voltages are maintained with a high degree of accuracy, so active rectifiers are excluded from the model. Thus, this model can be used to study the functioning of the electric drive properly, the rolling process, and the operation of inverters. To reduce the simulation time some more, the distances that the sheet passes between stands 1–2 and 2–3 are reduced in the model (to 4 and 6 m, respectively), although the actual inter-stand distance of 12 m is kept in the equations for tensions. Process is shown in Figure 1.31. It is seen that changes of the torques and tensions are the same practically, as in Figure 1.30, for the **Mill_rough2** model. Thus, this last, much simpler and faster model, can be successfully used to study the rolling process and the operation mode of the electric drive.

Mill_rough2a model is an expansion of the **Mill_rough2** model, in which the process of going out of the rolled sheet is simulated in addition. The ingot length is taken as 14.5 m, so that the rolling time in the first stand is $t_1 = 14.5 \times 124/(2.53 \times 79) = 9$ s. The circuits that fix sheet output from the next stand are added. So, for example, the

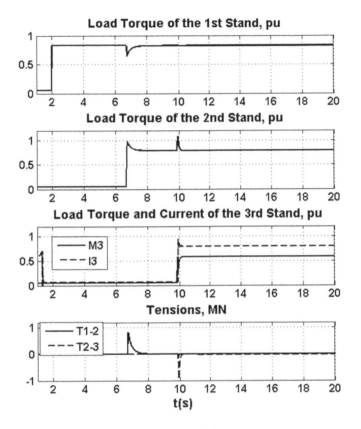

FIGURE 1.30 Processes in the model **Mill_rough2**.

bite2 signal is set, as before, after a time interval $t = 12/V_1$ after the sheet enters the first stand (which takes place at $t = 2\,\text{s}$); after the sheet leaves the first stand, the path traveled by the end of the sheet begins to be calculated as an integral $\Delta S = \dfrac{54}{79} \int V_2 dt$, and when this integral becomes 12, the *bite2* signal is reset. At this time, the calculation of the path traveled by the end of the sheet in front of the third stand as $\Delta S = \dfrac{40}{54} \int V_3 dt$ starts, and when this integral becomes equal to 12, the trigger **Bestable** is reset, which was set when the sheet entered the second stand, which leads to setting *bite3* to 0. The corresponding rolling cycle is shown in Figure 1.32, it lasts about 17 s.

The simplified models described can be used to study the rolling process itself, but they do not give an idea of the processes in active rectifiers and the network. To understand the processes in the grid supplying the roughing mill group drives, the active rectifiers are added to the **Mill_roug3** model in the **Mill_riugh3a** model. The models of these rectifiers are taken from the **Plate 34** model. Simulation in this model is carried out sometimes slower than in the **Mill_roug3** model. Figure 1.33 shows the grid current and powers consumed from the grid for the process that is

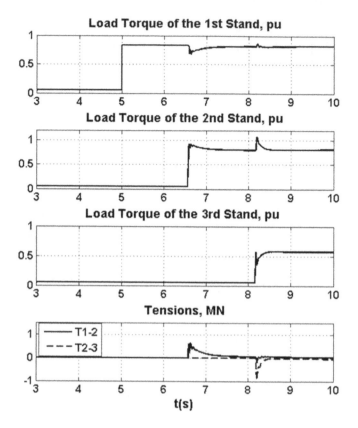

FIGURE 1.31   Processes in the model **Mill_rough3**.

shown in Figure 1.31. The total active power reaches 26 MW, reactive power is close to zero, and grid current THD is about 4%.

Let's consider one more point. Currently, much attention is paid to the development and application of multiphase electrical machines. This is of particular interest in relation to powerful machines, as it often makes it possible to provide greater fault tolerance and lower torque ripple, dispense with the parallel connection of inverters, which often causes certain technical difficulties. Therefore, the use of multi-phase, for example, six-phase synchronous motors for drives of mill stands seems to be very promising. Therefore, it seems appropriate to present in this book a model of such a motor, bearing in mind that, if necessary, it can be applied, with a corresponding change in parameters, for the simulation of new rolling mills. Such a model is given in the **Drive_12000_6Ph** model. Six-phase synchronous motor, with a power of 12 MW, has, as before, 10 pole pairs and a nominal frequency of 10 Hz, that is, a nominal rotational speed of 60 rpm. The remaining parameters are chosen, to a certain extent, at will, since there are no developments for such a motor or the author is not aware of it. The model of such a motor was developed by the author in Refs. [5,8] and takes into account flux linkages that couple both winding systems. The stator windings are powered by separate three-level inverters having a common DC link formed

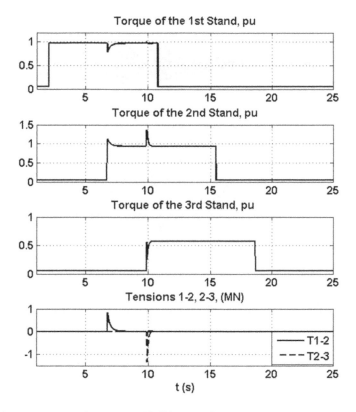

FIGURE 1.32 Processes in the model **Mill_rough2a**.

by the parallel connection of two three-level active rectifiers. The control system is the same as in the previous three-phase models, with the addition of circuits to form a second current regulator for the second winding, having an axis shift of 30° electrical. The process is shown in Figure 1.34 when the motor speeds up to the nominal speed of 60 rpm with an acceleration of 60 rpm/s, is loaded with the nominal load at $t = 3.5$ s, and speeds up with this load to 90 rpm.

1.2.2 FINISHING MILL GROUP DRIVES

We proceed to modeling the finishing group, which consists of seven stands (Figure 1.35), all stands are four-high mill stands, without a gearbox, Figure 1.5d. First, it is necessary to select electric motors, keeping in mind certain rolling schedules. Two schedules are considered. The first one: the sheet of 40 mm is rolled into a strip of 4 mm, width 1600 mm, with a rolling working speed of 12 m/s; the second schedule: the sheet of 32 mm is rolled into a strip of 2 mm, width 1250 mm, with a rolling working speed of 20 m/s. The diameter of the work rolls is 800 mm. The program $Sims1$ is used. The value of σ_s is calculated by (1.2). The results of calculations for the indicated two schedules are given in Tables 1.5 and 1.6 where T_m is the nominal torques of the selected motors at the given rotational speed (taking into account field reduction). The data of the selected motors are given in Table 1.7.

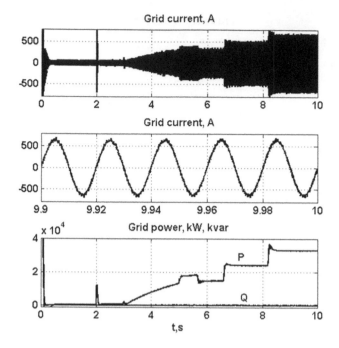

FIGURE 1.33   Grid current and powers for the roughing mill group drives.

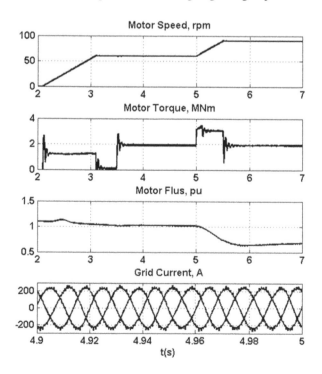

FIGURE 1.34   Processes in the six-phase synchronous motor.

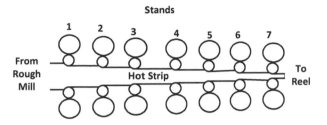

FIGURE 1.35   Seven-stand finishing group.

TABLE 1.5
Rolling 40–4 mm, Width 1600 mm

Stand	1	2	3	4	5	6	7
h_1, mm	24.7	13.2	9.3	6.9	5.5	4.5	4.0
σ_s, N/mm^2	97	126	135	146	150	167	154
T, °C	1080	1050	1000	980	960	920	900
P, MN	18.9	26.2	16.3	14.4	11.3	11.0	6.5
T_r, MN-m	1.1	1.19	0.51	0.36	0.23	0.19	0.085
N, rpm	46.4	86.8	123.2	166.1	208.3	254.6	286.5
T_m, MN-m	1.91	1.32	0.76	0.48	0.32	0.32	0.2

TABLE 1.6
Rolling 32–2 mm, Width 1250 mm

Stand	1	2	3	4	5	6	7
h_1, mm	14	7.2	5	3.8	2.9	2.3	2
σ_s, N/mm^2	124	148	156	166	189	205	193
T, °C	1060	1030	990	960	930	900	880
P, MN	25.7	22.8	13.4	10.7	11.4	10.5	6.5
T_r, MN-m	1.3	0.75	0.31	0.19	0.18	0.14	0.062
N, rpm	68.2	132.6	191	251.3	329	415.2	477.5
T_m, Mn-m	1.68	0.86	0.6	0.46	0.29	0.23	0.16

The motor torque available reserves are used to compensate for losses in motors and gears, to create a dynamic torque during the acceleration of the mill with metal in the rolls, to roll harder (alloyed) steels, to increase the rolling speed, which is a trend in modern rolling production. The powers of the selected motors correspond to those actually installed on modern rolling mills of similar types.

In the finishing group of stands, the strip should be rolled with a slight tension, approximately equal to 5–15 N/mm^2. To control the tension in the interstand spaces, loopers (loop holders) are installed, see Figure 1.36. With the help of a hydraulic (electric, pneumatic) drive, a certain torque is applied to the shaft of the

TABLE 1.7
Stand Motors

Stand	1, 2	3	4	5, 6	7
P, MW	12	12	12	10	8
N, rpm	60	150	250	300	400
T_m, MN-m	1.91	0.76	0.46	0.32	0.19
Z_p	10	10	10	8	6
f, Hz	10	25	41.7	40	40
H, s	0.24	0.7	1.3	1.7	2.5

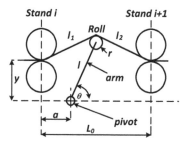

FIGURE 1.36 Looper.

looper lever, as a result of which a loop is created, the height of which is main-tained by adjusting the rotation speeds of adjacent stands. At a given loop height, the tension value is determined by the torque on the looper shaft, which is calcu-lated for given strip parameters in a given interstand space and for a wanted tension using known theoretical formulas (of course, with some error). In this way, indirect tension control is carried out.

Lately, the loopers have been used that directly measure the strip tension using strain gauges built into the body of the looper [15]. In this case, the torque applied to the looper shaft is controlled by the feedback principle, maintaining the reference tension value. In this case, there is also another possibility: to maintain the looper height by acting on the torque of the looper, and to regulate the tension value by act-ing on the rotation speeds of the stand motors.

Further, the construction parameters of the looper are taken, close to those given in Ref. [16]: $L_0 = 6$ m, $y = 0.19$ m, $a = 2$ m, r (radius) $= 0.15$ m, $l = 0.76$ m, lever mass 100 kg, roller mass 200 kg, moment of inertia 175 kg-m². In this case, the angle of contact of the looper roll with the strip (rolling level) is 2.9°.

The strip length in the interstand space $L_1 = l_1 + l_2$ is calculated by the following formulas [16]:

$$l_1 = \sqrt{(r-y+A)^2 + (a+B)^2}, \, l_2 = \sqrt{(r-y+A)^2 + (L_0-a-B)^2}, \quad (1.32)$$

$$A = l \sin\theta, \, B = l \cos\theta, \, C = A + r, \, D = B + a. \quad (1.33)$$

The tension equation receives a form

$$\frac{dT_{i,i+1}}{dt} = \frac{EQ_i}{L}\left(V_{in,i+1} - V_{out,i} + \frac{dL_1}{dt}\right) \tag{1.34}$$

During simulation, the last term can be obtained by differentiation of the given above expressions or by using the equality $\frac{dL_1}{dt} = \frac{\partial L_1}{\partial \theta}\frac{d\theta}{dt}$. The second multiplier is extracted from the looper drive circuit, and the first one is calculated using the relationships

$$\frac{dl_1}{d\theta} = l\left[(r-y)\cos\theta - a\sin\theta\right]/l_1 \tag{1.35}$$

$$\frac{dl_2}{d\theta} = l\left[(r-y)\cos\theta + (L_0-a)\sin\theta\right]/l_2 \tag{1.36}$$

Subsequent models often provide for both possibilities.

The equation of looper motion is written as

$$J_l\frac{d\omega_l}{dt} = M_l - M_t - M_{wp} - M_f, \frac{d\theta}{dt} = \omega_l \tag{1.37}$$

where J_l is the moment of inertia of the rotating parts of the looper, ω_l is its angular speed of rotation, M_l is the torque applied to the looper rotation shaft from its drive side, M_f is the torque of viscous friction, $M_f = k_f\,\omega_l$, M_t and M_{wp} are the torques applied to the looper roll and preventing its upward movement, namely [16]: M_t is the torque created by the tension of the strip T,

$$M_t = T\left[C\left(\frac{D}{l_1} - \frac{(L_0-D)}{l_2}\right) - B(C-y)\left(\frac{1}{l_1} + \frac{1}{l_2}\right)\right] \tag{1.38}$$

M_{wp} is the torque created by the weight of the strip in the interstand space and by the weight of the moving parts of the looper,

$$M_{wp} = (L_1Q_iG + G_l)B, \tag{1.39}$$

where G is the specific weight of the rolled material, G_l is the weight of the roll and half the weight of the lever.

The next few models are intended to acquaint with the basic principles of looper control.

In the **looper2** model, the operation of the looper between stands 3 and 4 is simulated when a sheet of 2 mm thickness is rolled. The distance between the stands is assumed to be 6 m. The parameters of the motors, stands, and strip are indicated above. It is assumed that the rolling conditions before the third stand and after the fourth stand remain constant. The parameters of the subsystems for the computation of the rolling torques and the values of the tension, in particular, the dependence of the forward slip on the tension, are changed in accordance with the above given calculations and relationships. In contrast to the roughing group, where the possibility of negative tensions is provided, in the finishing group, it is assumed that negative tensions are impossible: due to the small thickness, a loop will appear instead of negative tensions. The subsystem **dL_1/dt** is added in the subsystem **Tension**, in which, with two methods described above, calculations of the third term in (1.34) for $T_{i,i+1}$ and the value M_t by (1.38) are carried out. The circuits for the looper position control are in the subsystem **Load**, where the relationship (1.37) is carried out. Factor $k_f = 2$ kN/rad/s. It is assumed that the value of the torque M_{wp} is taken into account when calculating the reference torque of the looper, the *deltaM* value reflects the error in compensation. The torque reference is calculated using the same formula as the actual torque value for a given tension of 50 kN and an angle of 15°. The unit with the first order transfer function with a time constant of 50 ms models the looper torque control system.

Maintaining the position of the looper is carried out by action on the rotation speed of the third stand motor, that is, against the direction of the strip movement, using a PID controller.

The process of the strip input in the fourth stand and turning on the looper is shown in Figure. 1.37. It can be seen how, after the strip enters the stand, a slight tension arises in the strip, which, after turning on the looper, is set at a given level of 50 kN. At that, the circumferential speed of the third stand rolls is reduced from 8 to 7.96 m/s.

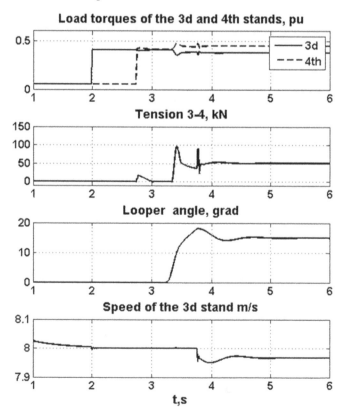

FIGURE 1.37 Strip input in the stand.

In the **looper2a** model, the response of the system to a perturbation is studied, which is taken as a change in the thickness at the exit of the third stand from 5 to 4.9 mm at $t = 6$ s. Since the rolling speed in the fourth stand and the thickness at its exit do not change, according to the condition, the speed of the strip entering the fourth stand increases and, as a result, the tension between the stands increases. Reduction of the output thickness also causes an increase of the rolling force in third stand, decrease of the rolling force in fourth stand, reduction of the cross-sectional area of the strip in the equation for changing the tension (1.34). Since the electric drives of the stands are equipped with PI speed controllers, speed changes caused by load changes are quickly counteracted, so they are assumed to have no noticeable effect on the tension change process, so, for simplicity, they are assumed to occur instantly, while for other factors a more realistic assumption is taken that the specified change in thickness occurs in 0.5 s.

The considered process is shown in Figure 1.38. It can be seen that the tension increases at first, which leads to a decrease in the loop, but as the rolling speed increases in the third stand, the looper position and the tension value return to their original values.

However, the rolling stands are equipped with strip thickness regulators, so that, for some time after the appearance of the disturbance, the thickness returns to the predetermined value. To simulate the thickness control process, the **Stand** subsystem

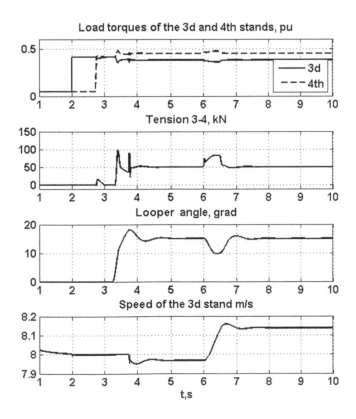

FIGURE 1.38   Process in the model **looper2a**.

was developed, see the **P_calc** model. It performs the same calculations as in the *Sims1* program, see also **Plate1a** model, **Rolling** subsystem. The evaluation of the strip thickness at the exit of the stand is made by relation (1.24). The input values are the strip thickness at the input h_{in} and the exit h_o of the stand, the strip temperature $T°C$, the rolling speed in the stand V, m/s, the rear T_i, and the forward T_0 of the strip tensions, N. The subsystem parameters box specifies the radius of the work rolls, the strip width, and coefficient K. The model provides the possibility to simulate the influence of the eccentricity of the rolls.

The subsystem makes it possible to calculate the forces and torques of rolling for given input and output thicknesses, but another problem arises during the rolling process: with known input thickness and position of the rolls, calculate the output thickness, taking into account the stiffness of the stand. To do this, we need to solve equation (1.24), bearing in mind the dependence of P on h_1. In the model under consideration, this is achieved by the fact that the signal $h_e = S_o + \dfrac{P}{K}$ is fed to the input h_o through the unit with the first-order transfer function that has a small time constant and the possibility to assign the initial value, which are taken equal to the assigned output thickness, as in the **Plate1a** model.

When studying the thickness control process, the position of the rolls changes in such a way that the value of h_e is equal to the wanted value of the output thickness. In the model under consideration, a proportional-integral controller is used for this purpose, and the reaction time of the hydraulic device for moving the rolls is assumed to be 30 ms, and the speed of moving the rolls is 2 mm/s.

To check the adequacy of the model, let's carry out the following calculations using the *Sims1* program.

Let $h_{in} = 7.2$ мм, $h_o = 5$ мм, rolling speed is 8 m/s. Then the rolling force $P = 13.4$ MN, rolling torque (without tension) $T_r = 0.31$ MN-m (see Table 1.6), and with $K = 6$ MN/mm $S_0 = 5–13.4/6 = 2.77$ mm. Let h_{in} reduce to 6.85 mm. Reiterating calculations for different values of h_o, it can be found that h_o is equal to h_e for $h_{out} = h_e = 4.86$ mm. At this, $P = 12.55$ MN, $T_r = 276$ kN-m. If to reach the wanted thickness $h_{out} = 5$ mm by the roll displacement, it will be $P = 11.41$ MN, $T_r = 247$ kN-m, and $S_0 = 5–11.41/6 = 3.1$ mm.

Operation of the thickness controller in response to reducing the input thickness from 7.2 to 6.85 mm at $t = 1$ s is shown in Figure 1.39 (**Switch2** tumbler is in the left position). The activation of the thickness control is delayed in relation to the moment of changing the thickness so that the new values had been reached the steady state. It can be seen that the values of thickness, rolling force, and torque correspond to the above given calculations. After turning on the controller, the value of the output thickness returns to its original value, and the values of the rolling force and torque correspond to the calculation, which confirms the adequacy of the developed model and the possibility of its use in modeling more complex processes in a rolling mill.

The eccentricity of the roll system is modeled in a simplified way, assuming that there is only the eccentricity of the backup rolls, which are identical [17]. In this case, (1.24) is replaced by $h_1 = S_o + \dfrac{P}{K} + S_e$, where the eccentricity is $S_e = E\sin\left(\dfrac{V_{out}}{R_b}t\right)$, where E is the eccentricity amplitude, R_b is the radius of the backup rolls. Let's carry

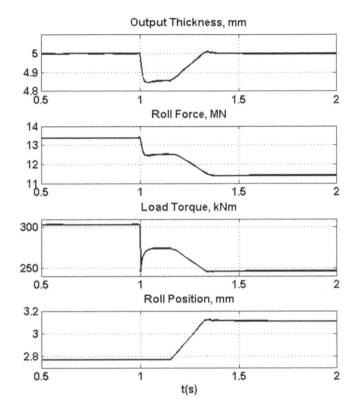

FIGURE 1.39 Stand model function.

out the simulation with $E = 0.05\,\text{mm}$, $R_b = 800\,\text{mm}$. **Switch2** in the right position. The input thickness is 7.2 mm and does not change, the thickness controller switches on at $t = 2.3\,\text{s}$. Process is shown in Figure 1.40. It can be seen that after turning on the thickness controller, the influence of the eccentricity is significantly reduced.

The **P_calc1** model simulates the passage of a perturbation in thickness through stands 3 and 4. The simulation is simplified, since changes in the interstand tension and load torques are not taken into account. The change in thickness at the exit of the third stand reaches the fourth stand in 6/8 s. It can be seen that the change in the input thickness of the third stand from 7.2 to 6.85 mm that occurs at $t = 2.5\,\text{s}$ is counteracted by the movement of the rolls of this stand, so that the input thickness of the fourth stand, after a short deviation, returns to the unperturbed value, as well as the position of the rolls of this stand, Figure 1.41.

The speed at the exit of continuous rolling mills is 20 m/s or more, while the threading speed of the strip into coiler is much lower. Thus, after the strip is threaded into the next coiler, the mill accelerates from threading to working speed. In the **Looper2b** model, the threading speed is equal to half the working speed (20 m/s) at the exit of stand 7. Process is shown in Figure 1.42. It can be seen that both looper position and the tension value, which were set during threading, are keeping up during acceleration and afterwards.

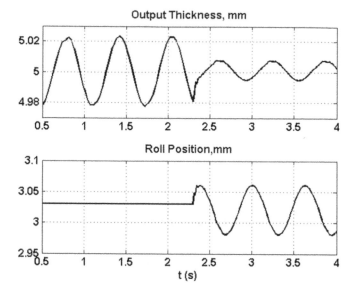

FIGURE 1.40 Back-up roll eccentricity affect.

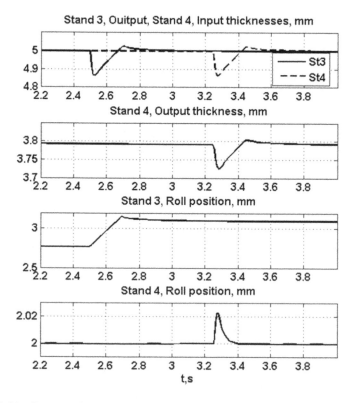

FIGURE 1.41 Passage of a perturbation in thickness through stands 3 and 4.

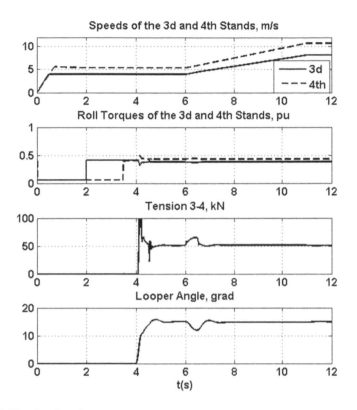

FIGURE 1.42   Acceleration process.

The **Looper2c** model is the **Looper2a** model, in which the **Stand** subsystems described above are added, in which the calculations of the forces and torques of rolling and the thickness of the strip are carried out. The change of the output thickness from the third stand reaches the fourth stand after 0.75 s. The calculated torques are used to determine the loads of the stand motors. Calculated (variable) values of strip thicknesses are used in tension calculations where these values are included in the corresponding relationships. The process is simulated when the strip enters the third stand at $t = 2$ s, into the fourth stand at $t = 2.75$ s with almost zero tension, after which the looper goes up with the given torque of 5 kN-m, and the specified tension of the strip between the stands of 50 kN reaches. At $t = 5$ s, the strip thickness at the input of stand 3 decreases from 7.2 to 6.85 mm at a rate of 10 mm/s, which leads to a decrease in the rolling torque in this stand, but the tension value does not change. After 0.75 s, the change in thickness reaches the fourth stand, which causes a short-term change in tension, which quickly returns to its original value, Figure 1.43.

The complete model of the 7-stand finishing group is presented in the **mill1_7** model. The model of each stand **Stand1—Stand7** contains the **Drive** subsystem, which simulates a motor with a speed controller, as in the **Looper2b, 2c** models, and the **Torque** subsystem, in which the load torque value is calculated, which is also similar to those used in the mentioned models. Before strip bite, thickness values are the design quantities that are used for mill adjustment, and after bite,

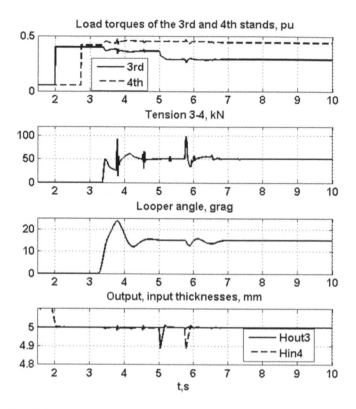

FIGURE 1.43　Thickness and tension control in the third and fourth stands.

they are the real values. The latter can differ from the former, since the value of the stand stiffness is known with some error, and the value of the rolling force can also be estimated with an error. The stand model provides for the possibility of simulating this phenomenon: the signal at the input h_o changes by 3% after the strip bite. Six subsystems **Interstand1_2—6_7** contain circuits for generating the strip tension signal between the specified stands and control circuits for loopers, also similar to those used in the **Looper2b, 2c** models. Blocks of variable delay are also placed here, simulating the passage of a change in the thickness of the strip between the stands; the delay values are defined as $\tau_{i,i+1} = L0/V_i$, where V_i is the speed of the strip leaving the i-th stand. In the **Nips** subsystem, signals *bite1—bite7* are generated, they are the instants when the strip enters the next stand. The forward slip values are calculated by the above given relations, with some correction due to the scarce experimental data that can be found in the literature.

Of course, when modeling a specific rolling mill, the parameters of drives, stands, loopers, screw-downs, and other equipment, which are rather arbitrarily taken in this model, must be replaced by parameters of the simulated mill, but, nevertheless, this model, which can serve as a basis for creating a custom model, makes it possible to get acquainted and better understand the main features of the functioning of a strip rolling mill.

When designing the control system, there are two possibilities for the action of the difference signal in the desired and actual position of the looper on the rotation speeds of the stand motors: on the previous stand or on the next stand. During threading the strip, it is obviously more reasonable to act on the next stand so that the already formed ratio of speeds in the previous stands does not change. However, when rolling, it is better to act on the previous stand in order to prevent the perturbation spreading to the exit stands. Since most of the strip is rolled at a steady speed, the second option is more often used, although other solutions are known. In the models under consideration, the action is carried out on the previous stand.

Work with this model shows that in the system under consideration there is a close link between state variables separated from each other by one or more stands. This circumstance makes it difficult to choose the parameters of control systems. It took a lot of time even to select the parameters used in the model, which, although they implement the required conditions, are not optimal. There are two approaches to solve this problem. In the first one, using empirical, intuitive grounds, connections are introduced that compensate, wholly or partly, for undesirable couplings in the system. The disadvantage of this approach is that it can improve performance with a particular combination of system parameters, and even worsen performance with another combination, and these results cannot be predicted in advance. In another approach, being based on the equations of the system, the behavior of the system as a whole is optimized in the presence of the abovementioned links. Thus, in a series of works by J. Pittner and M. A. Simaan, the finishing group is considered as a multi-variable system of 38 order, for the optimization of which the state-dependent Riccati equation (SDRE) based method is used, in which the Riccati equation with a new set of parameters must be solved anew on each discrete interval. This requires quite a lot of computational time, and in addition, to obtain consistent results, it is necessary that all parameters included in the matrix of equations refer to the same moment in time, which is not always easy to implement. Moreover, these methods are based on theoretical relationships describing the rolling process, the accuracy and adequacy of which are not always clear. Therefore, these methods are mainly of theoretical interest. References to almost all previous works of the authors are given in Ref. [18].

In the model under consideration, a serial action on the rotational speeds of the previous stands is carried out. The corrective actions are summed up and transmitted in the direction "against the rolling direction". In this case, the values of additional actions on the speed of the stands are carried out in the same percentage relative to the current speed of rotation of the stand motors.

The following values of the interstand tensions are taken: $T_{1-2} = 150$ kN, $T_{2-3} = 75$ kN, $T_{3-4} = 50$ kN, $T_{4-5} = 35$ kN, $T_{5-6} = T_{6-7} = 25$ kN. The process is shown in Figures 1.44 and 1.45 when the strip is threaded into the mill at a rolling speed of 10 m/s, and after setting the specified tensions, the mill speeds up to a speed of 20 m/s with an acceleration of 1 m/s². It can be seen that after the initial transient process, the values of the specified tensions and the positions of the loopers are maintained. As for the thickness of the strip after the stands, due to an error in the evaluation of the stiffness of the stand, instead of the given values of 14–7.2–5–3.8–2.9–2.3–2 mm, the values 14.4–7.4–5.15–3.91–3–2.37–2.06 mm are set, so that the use of an additional thickness regulator is required, which is considered in the following models.

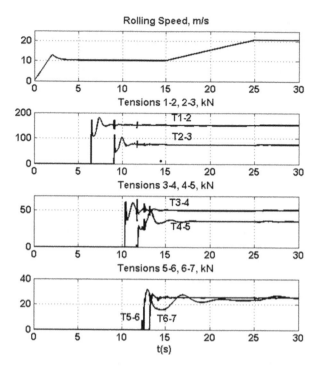

FIGURE 1.44   Tension changes during mill speeding up.

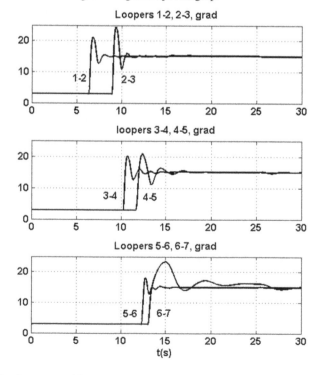

FIGURE 1.45   Looper position changes during mill speeding up.

The thickness of the strip at the exit of each stand is maintained equal to the specified value by means of a hydraulic device for moving the rolls using the relationship (1.24). Since the value of the stiffness K is not known exactly, there is some error in the thickness of the strip at the exit of the mill, and to compensate for it, a micrometer is installed after the last stand, with which the set point of the regulator of the last stand is adjusted. The micrometer is placed at some distance from the axis of the rolls of the last stand, so that there is a delay T_d between the thickness of the strip coming out of the stand of interest to us and the measured thickness. Two variants of the circuits with use of this thickness gauge are considered. In the first case, the deviation of the measured thickness from the wanted value is used directly to correct the reference thickness, Figure 1.46a. In the second variant, the micrometer measurements h_m are compared with the H_e thickness estimated by formula 1.24, corrected in such a way that the new value, shifted by T_d, and micrometer output h_m are the same, Figure 1.46b [16]. As can be seen from Figure 1.46a, before entering the comparison element, the value measured by the micrometer is averaged; Further, the averaging time is taken to be 0.2 s, which, for a rolling speed of 20 m/s, gives averaging over a section of 4 m length.

A simple **Thickness** model simulates only one, seventh stand and is intended for acquaintance and comparison of these options. The stand motor accelerates to a rolling speed of 20 m/s, at $t = 15$ s the strip enters the stand, and the value of h_o becomes equal not to the theoretical, but to the actual (taking into account the error in its calculation) value of the strip thickness leaving the stand; with the delay of 4 s, the correction circuit is switched on. At $t = 30$ s, the input thickness increases from 2.3 to 2.4 mm.

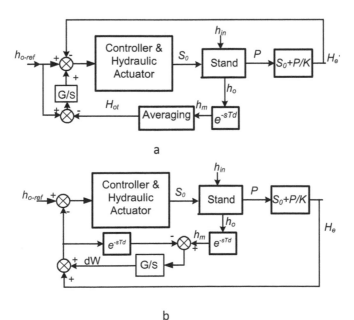

a

b

FIGURE 1.46 Thickness control structures, (a) correction reference value, (b) correction estimated value.

Figure 1.47 shows the simulation results. It can be seen that after the strip enters the stand, its actual thickness at the exit differs from the specified one, but after switching on the correction circuits, it acquires the wanted value of 2 mm. It can also be seen that the second method of the correction implementation, using a micrometer, provides faster convergence and less oscillation.

The **mill1_7b** model is a complete model of the finishing group, including errors in thickness measurement and the presence of a micrometer; the second thickness correction method mentioned above is used. It was found during simulation that the tension T_{6-7} is very sensitive to the change of the roll position of the seventh stand S_0, and in order for the tension changes to remain within acceptable limits, it is necessary to have a slow correction process, that is, the small value of the gain of the integrator G in Figure 1.46b, but at that the correction process is extremely slow. It was found that a significant increase in the value of G in the presence of relatively small tension fluctuations can be achieved by introducing an additional action of the signal dW from the integrator output on the rotation speed of the seventh stand motor. The process of threading, switching on the thickness corrector, acceleration, and operation at a steady speed of 20 m/s is shown in Figure 1.48.

Now the systems are considered, in which direct tension measurement is carried out. The most natural solution seems to be in which the control system of the looper maintains the set value of its angle, and the tension is controlled by action on the speed of rotation of the stand motors. A model for such a finishing group is given in the **mill1_7c** model, which assumes that the strip thickness is measured without errors. The block diagram of the control system is shown in Figure 1.49.

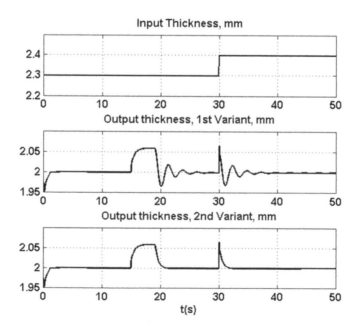

FIGURE 1.47 Strip thickness correction in the seventh stand.

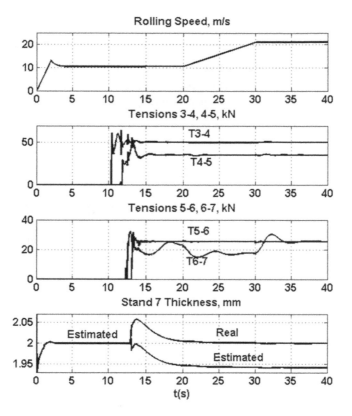

FIGURE 1.48 Threading, switching on the thickness corrector, acceleration, and operation at a steady speed.

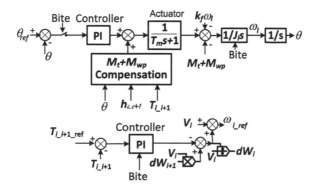

FIGURE 1.49 Block diagram control system with direct tension measurement.

If assume that the strip effects on the looper are compensated, the looper transfer function, using (1.37), can be written as

$$W_l = \frac{U^*}{sk_f(T_l s + 1)(T_m s + 1)}$$

(1.40)

where $T_l = J_l/k_f$, and T_m is a time constant of the looper drive control system taken, as before, equal to 50 ms. Taking the PI controller transfer function as $W_c = \dfrac{T_3 s + 1}{T_4 s}$, its parameters are determined depending on the selected maximum of the amplitude-frequency characteristic of a closed system M [19]. When M increases, the step response overshoot σ increases too, approximately

$$\sigma = \frac{M-1}{M}. \tag{1.41}$$

For the system (1.40)

$$T_3/T_\mu = (M+1)/(M-1), \, T_4/T_\mu = (M+1)^2 T_\mu / \left[k_f M (M-1) \right]. \tag{1.42}$$

Here $T_\mu = T_l + T_m$. We take M = 1.3 to have an overshoot about 25%. Also $T_\mu = 175/2000 + 0.05 = 0.138$. After calculations, it turns out $T_3 \approx 1.06$, $T_4 = 5.29 \times \dfrac{0.138^2}{2000 \times 1.3 \times 0.3} = 1.29 \times 10^{-4}$.

So, the PI controller parameters are $k_p = T_3/T_4 = 8217$, $k_i = 1/T_4 = 7519$. It was taken finally $k_p = 8000$, $k_i = 8000$.

The looper control system provides compensation for the load torque components caused by the weight of the strip and its tension in accordance with (1.38), (1.39); it is possible in this case, since the angle of the looper and the tension are measurable. The compensation error is assumed to be 10%. When calculating the compensating signal, it is still assumed that the strip thickness is equal to the nominal value.

In the **mill1_7c** model, the process of changing the interstand tensions is simulated when the strip is loaded into the mill, and the latter speeds up afterwards supposing an accurate measurement of the strip thickness in all stands and an error in compensating the looper load torque of 10%. Figure 1.50 shows the change in interstand tensions, and Figure 1.51 changes in the positions of the loopers. It can be seen that the process proceeds quite favorably.

The **mill1_7d** model assumes the existence of errors in thickness measurements, as in the **mill1_7b** model, and the presence of a strip thickness controller at the exit of the mill. As before, in the stand motor rotation speed control systems, the sequential actions on the rotation speeds of the previous stands is carried out; the corrective actions are summed up and transmitted in the direction "against the rolling direction". The magnitudes of additional actions on the speeds of the stands are carried out in the same percentage with respect to the current rotation speed of the stand motors. The processes in the system are approximately the same as in the **mill1_7b** model and are not shown here.

The following models simulate an electric motor and its power and control system in detail. Unfortunately, simulation in such models proceeds extremely slowly; therefore, simulation using such models is performed fragmentarily.

Simulation of two adjacent stands is carried out in the **mill_finish2_3**, **mill_finish4_5**, **mill_finish6_7 models**. The stands with the motors having the different rated data are chosen. The same systems are employed for their power supply and

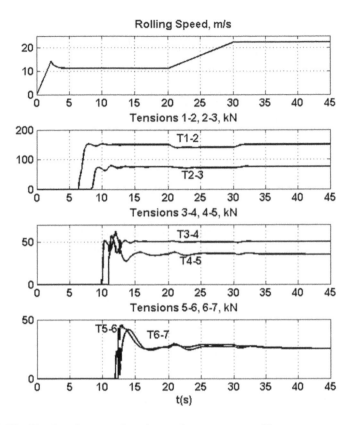

FIGURE 1.50 Tension changes when the tensions are measurable.

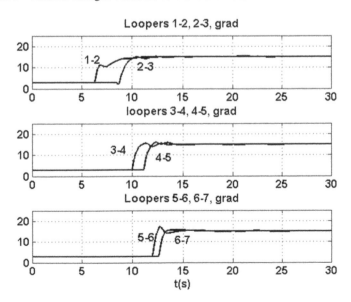

FIGURE 1.51 Looper position changes when the tensions are measurable.

control as for the roughing group: with three-level voltage inverters. The interstand span models are the same as in the **mill1_7** model. From there, regulators of interstand tensions are borrowed, which affect the speed of rotation of the previous stand motor. Since the back tension for this stand and the forward tension for the next stand are not modeled, it is assumed that after the strip enters the first stand, its back tension equal to the specified value appears, and after the strip enters the stand following the second, the forward tension of the second stand appears, also equal to the specified value.

The stand models provide the possibility to simulate the error in the estimation of the strip thickness at the exit of the stand.

For the motors with power of 12,000 kW, the same power supply system is used as for the motor of the same power of a plate mill, Figures 1.25 or 1.27. It is assumed in these models that the DC voltages at the input of the inverters are maintained with a high degree of accuracy, so that the input active rectifiers are not modeled. The motors speed up from null to the threading speed that is 10 m/s at the group output, at $t = 5$ s the strip enters the first of two stands, after time $6/V_2$ ($6/V_4$ или $6/V_6$) in the second stand, and after time $6/V_3$ or $6/V_5$ in the next stand (which is not modeled, for the seventh stand this latter condition is not modeled), whereupon, at $t = 8$ s, mill speeding up to the working speed of 20 m/s begins; at this, to reduce simulation time, the acceleration is increased by two times. So, the motor dynamic torque is increased twice. For stands 2–3, the process is shown in Figure 1.52. The decrease of the motor flux linkages is

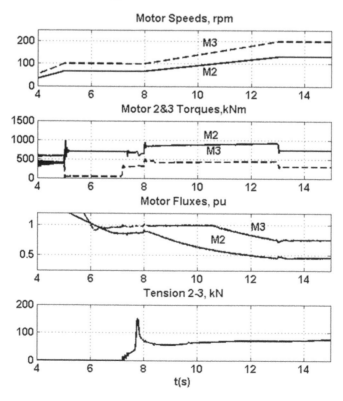

FIGURE 1.52   Rolling process in stands 2 and 3.

clearly visible when their rotation speeds begin to exceed the nominal values. A number of oscilloscopes placed in the model make it possible to observe changes in many system variables. It can be noted that the acceleration section from 0 to the threading speed is of no practical importance, since it occurs without metal and is extremely rare.

If to add an active rectifier to the considered model, then the simulation will slow down more. At the same time, the processes in the motor itself, in its control and power systems, and in the rolling mill proceed in this model in the same way as in models without modeling active rectifiers, and additional information consists in the effect of the electric drive on the supply network. For this purpose, it is sufficient to be able to simulate the electric drive of each individual stand in order to obtain a faster estimate of the reactive power of the network and the shape of the current in it.

The first or second stand is simulated in the **Finish_Mill_Drive_12MW_60** model. The power supply and control systems are the same as in **Plate 32** model. The motor accelerates to the threading speed (corresponding in this case to 10 m/s for the seventh stand), without a load; at $t = 6$ s, the strip enters the stand, and the motor receives a load corresponding to the rolling process; at $t = 8$ s, the motor begins to speed up to the operating speed (corresponding in this case to 20 m/s for the seventh stand), with an acceleration of 0.125 pu/s. The process for the second stand is shown in Figure 1.53. The network reactive power is nearly null; it can be

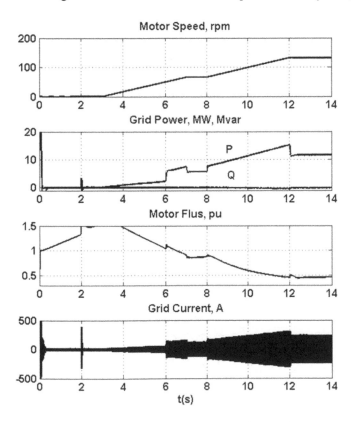

FIGURE 1.53 Process in the model **Finish_Mill_Drive_12MW_60**.

seen in the considerable reduction of the stator linkage when speeding up to the operating speed. THD in network current is about 3% at $t = 12$ s, approximately the same THD value is in the motor current. Similar results take place for the third stand, **Finish_Mill_Drive_12MW_150** model. For the fourth stand, **Finish_Mill_Drive_12MW_250** model, THD in the network current is some more, 4%–4.5%.

For stands 5–7, due to a decrease in motor power, instead of three parallel-connected rectifier-inverter circuits, two circuits of this type are used, connected to the network through a three-winding transformer, the secondary windings of which have delta and star connection, Figure 1.54. For fifth and sixth stands, this model is **Finish_Mill_Drive_10MW_300**. Figure 1.55 shows the process for the sixth stand when rolling a 2 mm strip. The reactive power is close to zero, and the THD in the main current is less than 2%. Approximately the same results are obtained when modeling the seventh stand, model **Finish_Mill_Drive_8MW_400**.

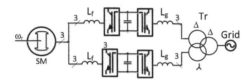

FIGURE 1.54   Supply power circuits for the motor 10 MW.

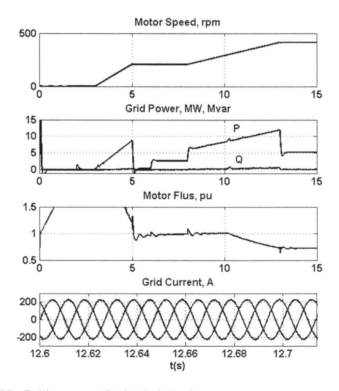

FIGURE 1.55   Rolling process for the sixth stand.

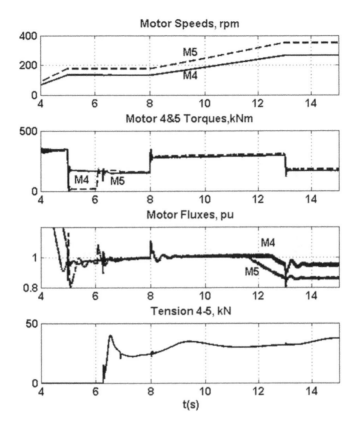

FIGURE 1.56 Rolling process in stands 4 and 5.

As it has been said earlier, the acceleration section from 0 to the threading speed is of no practical importance, and the values of acceleration are set larger than in reality.

The **mill_finish4_5** model is a revision of the **mill_finish2_3** model for stands 4 and 5. The process is shown in Figure 1.56. The **mill_finish6_7** model simulates the operation of stands 6 and 7. It takes into account the possible inaccuracy of indirect thickness measurement and the presence of a thickness regulator with a micrometer. The process is shown in Figure 1.57.

1.2.3 ELECTRIC DRIVE OF FLYING SHEARS

There are a number of auxiliary mechanisms in the line of a continuous hot rolling mill, from which one can distinguish the flying shears before the finishing group and the coilers after it. Flying shears are intended for cutting metal on the run ("on the flying "). Coiler drive simulation is considered in the section on cold rolling mills. Here we will focus on the simulation of the electric drive of the shears intended for cutting off the front end of the strip, which usually has defects, with a length of 150–250 mm. These are usually double-drum flying shears. The cutting process is schematically shown in Figure 1.58.

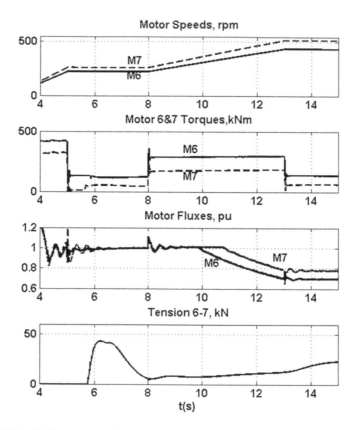

FIGURE 1.57 Rolling process in stands 6 and 7.

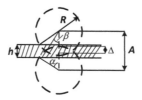

FIGURE 1.58 Strip cutting scheme.

At first, one must be able to calculate the cutting torque. Further, a cut of the strip with a size of $40 \times 1600\,\text{mm}$ ($h \times b$), the speed of the strip during the cut is $2.8\,\text{m/s}$, the diameter of the circle of the knives movement $D = 2R = 1120\,\text{mm}$.

The value of the ultimate strength of the material at the cutting temperature is taken equal to $\sigma_b = 80\,\text{N/mm}^2$, knife overlap $\Delta = 1\,\text{mm}$. The detachment of the metal cutting-off part occurs at a relative notch ε, the value of which is taken equal to $\varepsilon = 0.7$. Then the notch depth $y = 0.5 \times ((1-\varepsilon)h + \Delta) = 0.5 \times (0.3 \times 40 + 1) = 6.5$ mm.

The angle α at which the knives come into contact with the strip is equal to

$$\alpha = arccos\frac{D - h - \Delta}{D} = arccos\frac{1120 - 41}{1120} = 15.55° = 0.271\,\text{rad}.$$

Shear exit angle after cutting is $\beta = arccos\dfrac{0.5D - y}{0.5D} = 8.7 = 0.15$ rad.

Maximum cutting force [20]

$$P_s = K_1 K_2 \sigma_b h b (1 - \varepsilon_d),\tag{1.43}$$

where the factor K_1 takes into account the decrease in cutting resistance in relation to the value of ultimate strength and is taken equal to $K_1 = 0.7$, the factor K_2 takes into account additional resistance to cutting due to wear of the knives and efforts to move forward the material being cut, it is taken $K_2 = 1.25$, the quantity ε_d is the so-called dent-factor, which depends on the properties of the material and temperature and is taken equal to 0.3. Then $P_s = 1.25 \times 0.7 \times 80 \times 40 \times 1600 \times 0.7 = 3.1$MN.

The arm of force $c = 0.5D \times \sin \alpha = 150$mm, the torque $T = 3.1 \times 0.15 = 0.47$ MN-m.

Taking the cutting speed 15% more than the strip speed, the cutting rotational speed is $2.8 \times 1.15/0.56 = 5.75$ rad/s, that is, 55 rpm. The mechanism moment of inertia (without motor) is taken equal to 17,360 kg-m².

The ABB asynchronous motor with the main parameters of 3 kV, 50 Hz, $Z_p = 6$, that is, with the synchronous speed of 500 rpm, is chosen to drive the considered flying shears. The gear ratio is 9. The required motor torque is $470/9 = 52.2$ kN-m. The motor has parameters 1800 kW, 494 rpm, $\cos \varphi = 0.81$, $I_{nom} = 445$ A, $T_{nom} = 34.8$ kN-m, $J = 222$ kg-m², the maximal torque 2.1 pu-, no-load current 181 A.

Let's check the dynamic properties of the selected motor. Suppose that from the zero position to the cut, the shears should rotate by $190° = 3.32$ rad and at the same time reach the specified rotation speed of 5.75 rad/s. Then the acceleration should be $\dfrac{d\omega_s}{dt} = \dfrac{5.75^2}{2 \times 3.32} = 5$ rad/s², which, for the motor shaft, is $\dfrac{d\omega_r}{dt} = 45$ rad/s². The total reduced moment of inertia $J = 222 + 17,360/81 = 436$ kg-m². To create such acceleration, a dynamic torque of $436 \times 45 = 19,620$ N-m is required, which the selected motor provides, taking in mind that a cut and a speeding up are separated in time.

To power and control the shear motor, a DTC system for the asynchronous motor with a three-level voltage inverter is used. Such a system was studied in Refs. [5] and [21] and is implemented in the **DTC_shear1** model. The knives starting position is taken equal to $3.6 \times 57.3° = 206°$, taking as null the low point, Figure 1.59.

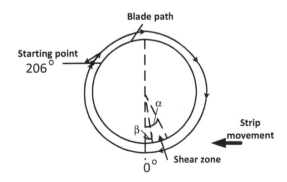

FIGURE 1.59 Knives movement plan.

The acceleration value is taken equal to 1.5 pu, and deceleration 2 pu. Because $1\,\text{pu} = 2 \times \pi \times 50/Z_p/i = 314/(6 \times 9) = 5.815\,\text{rad}$, value of acceleration is $a_a = 8.72\,\text{rad/s}^2$, and deceleration $a_d = 11.62\,\text{rad/s}^2$. Let's find time from the start to the instant of cut when the strip moves with the speed V_m. The speeding up time to the cut speed is $t_a = \dfrac{1.15V_m}{0.5Da_a} = \dfrac{1.15 \times 2.8}{0.56 \times 8.72} = 0.66$ s. At this, the knives will turn to the angle

$S_a = \dfrac{(1.15 \times 2.8)^2}{2 \times 8.72 \times 0.56^2} = 1.9$ rad $= 108.9°$. The angle α, when the knives come into contact with the strip, was calculated earlier and equal to 0.27 rad. So, at the steady speed, the knives have to turn by $206 - 108.9 - 15.5 = 81.6° = 1.43\,\text{rad}$ that takes $t_{ss} = \dfrac{1.43 \times 0.56}{1.15 \times 2.8} = 0.25$ s. Therefore, the full time from start to cut is $t_s = 0.91$ s.

Suppose that the signal for the approach of the strip comes from a photoelectric relay or another device located at a distance $L_f = 4\,\text{m}$ from the axis of the shears, and the length of the cut piece is $L_s = 0.2\,\text{m}$. Then the time remaining before the cut is $t_m = (L_f + L_s - \int V_m dt)/V_m$. When the current value of t_m becomes equal to the previously calculated value of t_s, the shear motor is turned on to perform the cut.

In the **DTC_shear1** model, the cutting process control circuits are placed in the **Shear_Control** subsystem. It calculates the values of t_m and t_s, according to the above given relationships, and it is possible to increase t_s value by some small amount to compensate for unaccounted delays (taken as 40 ms). When these times become equal, **Bistable** and **Bistable1** triggers are set, *On* and *On1* signals are generated, and a reference signal, which is proportional to the strip speed, is applied to the speed controller input. The motor is speeding up, the knives approach the strip. The current position of the knives is fixed by the integrator, which integrates the value of the angular speed of the shear drum in the direction of decreasing position, since the position of the knives at the bottom point is taken as zero. The initial value of the integrator output S_0 is equal to the abovementioned value of 3.6 rad. When the integrator output S becomes equal to the cut start angle α, the motor torque increases to the value of the cut torque calculated above, and when the integrator output S becomes equal to the cut end angle β, the torque decreases to the value of the idle torque. The **Bistable1** trigger is reset, *On1* = 0, in 0.2 s after the knives have passed the bottom position, In this case, the motor rotation speed is determined by the output of the proportional position controller with output limitation, which is assumed to be ± 0.2 pu. Let's pay attention that with the accepted scheme of calculations, the required position is equal to $S_0 - 2\pi$. As can be seen from Figure 1.60, the angle remaining from the end of the cut to the initial position is not sufficient to stop rotation of the shears; the knives miss the set position. Then they return to the desired position and the motor stops. After it stops, the **Bistable** trigger is reset, *On* = 0, and the initial value of S_0 is loaded into the integrator. The system is ready for the next start. This subsystem also contains circuits for controlling the plotter showing the movement of the knives.

The active rectifier circuit based on a three-level inverter is borrowed from previous models. The voltage in the DC link is assumed to be 5 kV and the transformer secondary voltage is 3.2 kV.

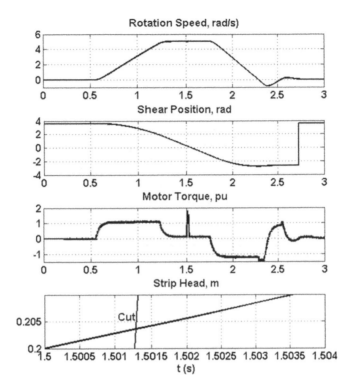

FIGURE 1.60 Process of the front end cut at strip speed 2.8 m/s.

Figure 1.60 shows the process at a strip speed of 2.8 m/s, and Figure 1.61 at 1.4 m/s. As can be seen from the last graphs of these figures, where the moment of the beginning of the cut is fixed, the deviation of the length of the cut piece from the wanted one (200 mm) does not exceed a few mm and, in principle, can be compensated. Changes in the current, active, and reactive power of the network are shown in Figure 1.62. It can be seen that the current is almost sinusoidal, and the reactive power is close to zero.

To finish modeling flying shears, their employment to cut material into the sections of the given length is considered. For specific data, let us consider a cut of a bar with a size of 100×100 mm, emerging from the last stand of the bar roll mill with a speed of 1.5–3.5 m/s, into sections of 4–12 m long, the diameter of the trajectory of the knives, as before, $D = 1120$ mm.

Ultimate strength of the material at a temperature of 700°C $\sigma_b = 150$ N/mm², relative notch $\varepsilon = 0.35$, knife overlap 1 mm, notch depth

$$y = 0.5 \times \left((1-\varepsilon)h + 1 \right) = 0.5 \times (0.65 \times 100 + 1) = 33 \text{ mm}.$$

The angle α at which the knives come into contact with the strip is equal to

$$\alpha = arccos \frac{D - h - \Delta}{D} = arccos \frac{1120 - 101}{1120} = 24.5° = 0.427 \text{ rad}.$$

Shear exit angle after cutting is $\beta = arccos \dfrac{527}{560} = 19.78 = 0.345 \text{ rad}.$

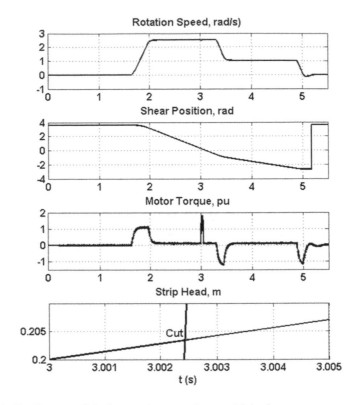

FIGURE 1.61   Process of the front end cut at strip speed 1.4 m/s.

Taking $K_1 = 0.74$, $K_2 = 1.3$, $\varepsilon_d = 0.2$, we find by (1.43) $P_s = 0.74 \times 1.3 \times 150 \times 100 \times 100 \times 0.8 = 1.15\,\text{MN}$.

The force arm is $0.5 \times 1120 \times \sin 0.427 = 232\,\text{mm}$, then the torque is $T = 1.15 \times 0.232 = 0.27\,\text{MN-m}$.

Taking the cut speed of 15% more than the bar speed, the rotational speed of the knives during a cut is $3.5 \times 1.15/0.56 = 8.05\,\text{rad/s}$, that is, 77 rpm.

The motor with $Z_p = 5$ (synchronous rotational speed 600 rpm) is selected, and the gear ratio is taken $i = 7.5$. Then the load torque reduced to the motor shaft is $266/7.5 = 35.47\,\text{kN-m}$. The proper motor has data 2240 kW, 593 rpm, $\cos\varphi = 0.85$, $I_{nom} = 524$ A, $T_{nom} = 36.04$ kN-m, $J = 215.7$ kg-m², maximal torque $T_{max} = 1.9$ pu, no-load current 147 A.

The knives initial position is taken, as before, equal to $3.6 \times 57.3° = 206°$. The acceleration value is taken equal to 1.5 pu, deceleration 1.8 pu. Because 1 pu is equal to $2 \times \pi \times 50/Z_p/i = 314/(5 \times 7.5) = 8.373\,\text{rad}$, the acceleration $a_a = 12.56\,\text{rad/s}^2$, deceleration $a_d = 15.07\,\text{rad/s}^2$. The time from the knives starts to the cut, when the metal moves with the speed V_m, $t_s = t_a + t_{ss}$, where t_a is the speeding up time to the cut speed, $t_a = \dfrac{1.15 V_m}{0.5 D a_a} = \dfrac{1.15 \times 3.5}{0.56 \times 12.56} = 0.57\,\text{s}$. At this, the knives turn to the angle

$$S_a = \frac{(1.15 \times 3.5)^2}{2 \times 12.56 \times 0.56^2} = 2.056\,\text{rad} = 117.8°. \text{ The angle } \alpha, \text{ when the knives come}$$

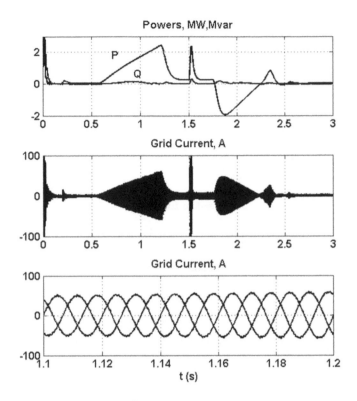

FIGURE 1.62 Network current and power at cut.

into contact with the metal, was calculated earlier and equal to 24.5°. So, at the steady speed, the knives have to turn by $206 - 117.8 - 24.5 = 63.7° = 1.11$ rad that takes $t_{ss} = \dfrac{1.11 \times 0.56}{1.15 \times 3.5} = 0.15$ s. Therefore, the full time from start to cut is $t_s = 0.72$ s.

If, after cutting off the front end, the shears continue to rotate at the same speed, then the strip will be cut into sections with a length of $(2 \times \pi \times 0.15)/1.11 \times 3.5 = 3.06$ m, while the minimum length of the section is 4 m. Figure 1.63 shows the movement of the metal being cut (named Strip in the figure) and plots of the motor torque, which show the instants of the cut. For the convenience of viewing the plots, the values of the motor torques are increased so that they are commensurate with the distance traveled by the metal. Figure 1.63a shows a cut with a constant rotational speed of knives.

To increase the time between cuts at a constant cutting speed, it is possible to reduce the speed of rotation of the knives between cuts. Let us assume that immediately after the cut, the rotational speed decreases during the time t_d with the accepted deceleration a_d, after which the rotational speed increases to the initial value V_N with acceleration a_a for the time t_a. We have $V_1 = V_N - a_d t_d$, $S_1 = V_N t_d - 0.5 a_d t_d^2$, $V_2 = V_1 + a_a t_a$, $t_a = a_d t_d / a_a$, $S_2 = V_1 t_a + 0.5 a_a t_a^2 = a_d t_d / a_a \times (V_N - 0.5 a_d t_d)$, so that, during slowing down and speeding up, the path will be covered

$$S_3 = (1 + a_d / a_a) \times t_d \times (V_N - 0.5 a_d \, t_d). \tag{1.44}$$

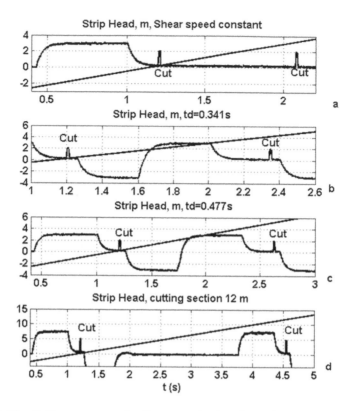

FIGURE 1.63 Cutting processes for diverse section lengths (a) 3.06 m; (b) 4 m; (c) 4.89 m; (d) 12 m.

The time interval between cuts will be equal to

$$t_{sr} = \frac{2\pi - S_3}{V_N} + t_d\left(1 + \frac{a_d}{a_a}\right),$$ (1.45)

and the length of cut sections $S_m = V_m t_{sr}$. To find t_d for the given value S_m the equation (1.46) is received

$$2\pi - \frac{V_N S_m}{V_m} + AV_N t_d = At_d\left(V_N - 0.5a_d t_d\right)$$ (1.46)

where $A = 1 + \dfrac{a_d}{a_a}$. Or

$$t_d = \sqrt{\frac{\dfrac{V_N S_m}{V_m} - 2\pi}{0.5Aa_d}}.$$ (1.47)

Let $S_m = 4$ m. Substituting into (1.47) the quantities $V_m = 3.5$ m/s $V_N = 1.15 \times 3.5/0.56 = 7.19$ rad/s, $a_d = 15.07$ rad/s², $A = 2.2$, we receive $t_d = 0.341$ s. Process is shown in Figure 1.63b.

The maximal time t_d is $t_{dmax} = V_N/a_d$. Therefore, for slowing down and speeding up, the path (in rad) will be covered

$$S_3 = A \frac{V_N^2}{2a_d}. \tag{1.48}$$

It follows from (1.47)

$$S_m = \frac{V_m}{V_N}\left(2\pi + \frac{AV_N^2}{2a_d}\right) \tag{1.49}$$

In the case under consideration, $t_{dmax} = 7.19/15.07 = 0.477$ s, $S_m = 4.89$ m, process is shown in Figure 1.63c.

When longer bar sections are needed, the motor works with stops and following starts. The stop duration is $t_p = \dfrac{S_m^* - 4.89}{V_m}$, где S_m^* is the required length of the cut section. At $S_m^* = 12$ m, the cutting process is shown in Figure 1.63d. The total deceleration and pause time $t_q = t_d + t_p = 2.51$ s.

Let $V_m = 1.5$ m/s and correspondingly $V_N = 3.08$ rad/s. When the knives rotate with a constant speed, the cut section length is the same, 3.06 m, but $t_{dmax} = 3.08/15.07 = 0.204$ s, so that the maximal cut length without pauses is $S_m = \dfrac{1.5}{3.08}\left(2\pi + \dfrac{1.1 \times 3.08^2}{15.07}\right) = 3.4$ m, Figure 1.64a. When the length of the cut sections is 12 m, $t_p = 5.74$ s, so that $t_q = 5.94$ s, Figure 1.64b.

The given figures are received with the **DTC_Shear2** model. The power supply and motor control systems are the same, as in the **DTC_Shear1** model. The circuits for control of the speed and position of the knives are placed in the subsystem **Shear_Control**. The circuits for the shear start to cut the front end are located in the subsystem **Start**; it was described for the **DTC_Shear1** model. When the *Start* signal appears, both triggers **Bistable, Bistable1** are set. The first one is reset after the last cut and resetting of the knives. On the model, the corresponding circuits are disconnected. Turning on the second trigger leads to setting the motor rotation speed, proportional to the metal speed. After passing the zero position, this trigger is reset; the motor starts to slow down. This turns on the count of the deceleration time with the help of the integrator. After termination of the slowing down time plus the stop time, the trigger is set and gives the motor speed. The t_d calculation is carried out according to the relationships given above. The calculated time t_d is compared with its possible maximum value t_{dmax}; if the first value is less than the second, then it is used to determine the instant when the trigger is set, and if it is greater than the second, then the motor must work with stops, and the total time of the reset state of the trigger is calculated as $t_{dmax} + t_p$.

Some graphs of processes when cutting to lengths of 4 and 8 m are shown in Figures 1.65 and 1.66, respectively. One can see that in the first case, the shear motor operates continuously, and in the second one with stops.

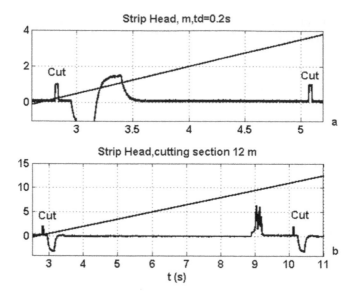

FIGURE 1.64   Cutting processes at the bar speed of 1.5 m/s. (a) Cut without stops. (b) Section length 12 m/s.

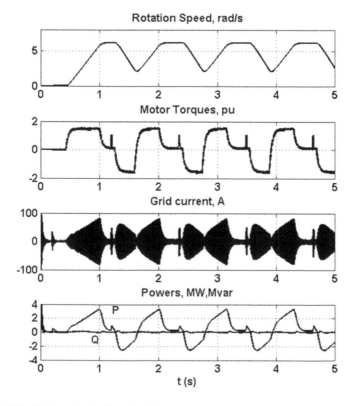

FIGURE 1.65   Cutting for the length of 4 m.

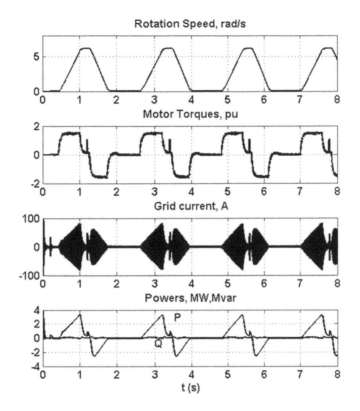

FIGURE 1.66 Cutting for the length of 8 m.

1.3 ELECTRICAL DRIVES OF THE COLD ROLLING MILLS

1.3.1 ELECTRICAL DRIVES OF THE REELS

At a cold rolling mill, rolling begins with the unwinding of the strip supplied in coils and ends with the winding of the finished strip into a coil. Therefore, it is reasonable to start the study of such mills with the simulation of the electric drive of the reels. The output manufacturing at continuous hot rolling mills, discussed in the previous paragraph, also ends with the winding of the strip into coils, the electric drives of the reels of both types of mills are largely identical.

As an example, the reel (the another term "coiler" will be used too) that operates in a one-stand reversing cold rolling mill is considered which rolling a strip of 1610 mm width; the mass of the coil is 36 t, the reel drum diameter is 0.6 m, the coil maximal diameter is 2 m, the rolling speed is 10 m/s. The strip comes to the mill with a thickness of 6 mm and is rolled into the strip with a thickness of 1.6 mm in five passes, with thicknesses after the passes: 6–4.38–3.11–2.4–1.8–1.6 mm. In such a mill, coilers are installed on both sides of the stand, in each pass one of them operates in the winding mode, and the other in unwinding mode.

The tension during winding is approximately 10% of the yield strength; by the calculation of which it is necessary to take into account the strain hardening of the material, that is, its hardening under deformation. This phenomenon is discussed in more detail in the next section. Here we only note that for the low-carbon steel, we can take the yield strength σ_f after the i-th pass equal to

$$\sigma_f = 230 + 34.6 \times \left(0.67 \times 100 \times \frac{h_0 - h_i}{h_0} \right)^{0.6} \text{N/mm}^2, \tag{1.50}$$

where h_0 is the thickness of the strip entering the mill without hardening, $h_0 = 6\,\text{mm}$ in our case, h_i is the thickness of the strip after the i-th pass. Then for the first pass

$(h_1 = 4.38\,\text{mm})$ $\sigma_f = 230 + 34.6 \times \left(0.67 \times 100 \times \frac{1.62}{6} \right)^{0.6} = 427\,\text{N/mm}^2$, for the last

pass $(h_5 = 1.6\,\text{mm})$ $\sigma_f = 230 + 34.6 \times \left(0.67 \times 100 \times \frac{4.4}{6} \right)^{0.6} = 588\,\text{N/mm}^2$.

The specific tension σ_r is taken no more than 60 N/mm², the full tension T no more than $T_{max} = 100$ kN. The drum maximal rotational speed $\omega_r = \frac{10}{0.3} = 33.3$ rad/s or $N = 318$ rpm. To distinguish between tension and torque, the latter will be denoted here as M. The motor torque at this speed is $M = 100,000 \times 0.3 = 30\,\text{kN-m}$; for the coil of the maximal diameter, the rotational speed is $33.3 \times 0.6/2 = 10\,\text{rad/s} = 96\,\text{rpm}$, $M = 100$ kN-m. The synchronous motor 2650 kW, 250 rpm, $Z_p = 10$, the nominal frequency of 41.7 Hz is chosen, the drive without gear $(i = 1)$.

When vector control principle is used, the I_q component reference must be $I_q = \frac{TD}{2\Psi i}$, where T is the wanted tension value, D is the coil diameter, Ψ is the stator linkage. Because the model operates in pu, pu tension units are introduced. One can write

$$I_q^* = \frac{TDU_b}{2P_b \Psi^* \Psi_b i} = \frac{TD}{2\Psi^* M_b i} \tag{1.51}$$

It is accepted $T_b = \frac{2M_b i}{D_{max}}$, and further $T^* = T/T_b$, $D^* = D/D_{max}$. Then

$$I_q^* = \frac{T^* D^*}{\Psi^*} \tag{1.52}$$

When winding begins, $\Psi^* = 250/318 = 0.786$ and $D* = 0.3$, so that for $T^* = 1$ $I_q^* = 0.38$. When the coil diameter increases, the rotational speed reduces and the value Ψ^* increases inversely. When the diameter increases to 0.38, the value Ψ^* becomes equal to 1 and does not change with a further increase in diameter. To maintain the tension given value, the quantity I_q^* has to increase in proportion to the increase in diameter.

Moments of inertia of the motor and drum are equal $J_{motor} = 2500$ kg-m² $(H_{motor} = 0.32\,\text{s})$, $J_{dr} = 1200$ kg-m² $(H_{dr} = 0.15\,\text{s})$ respectively, that is, the device total moment of inertia is $J_{mex} = 3700$ kg-m^2 ($H_{dr} = 0.47$ s). The coil moment of inertia reduced to the motor shaft is $= \dfrac{\pi\left(D^4 - D_0^4\right)b \times 7800}{32i^2}$, its maximal value is

$$J_{max} = \frac{\pi\left(2^4 - 0.6^4\right) \times 1.61 \times 7800}{32} = 19,566 \text{ kg-m}^2 \ (H_{dr} = 2.5 \text{ s}).$$

Thus, when winding, the moment of inertia of the motor with the attached load changes, which is not provided by the standard model of the synchronous motor.

One can write

$$\left(J_{mex} + J\right)\frac{d\omega_m}{dt} = M_m - T\frac{D}{2i} - M_{loss} \tag{1.53}$$

where M_m is the motor torque, M_{loss} is the loss torque. Or

$$J_{mex}\frac{d\omega_m}{dt} = M_m - T\frac{D}{2i} - J\frac{d\omega_m}{dt} - M_{loss} = M_m - M_{sh} \tag{1.54}$$

It should be noted that changes have been made to the previously used structure of the speed controller, providing for the possibility of externally setting the values of the controller output limits. We also note that in the motor control system, the **Vector Control** subsystem, there are two options for assigning the angle, according to which the armature current components $I_T = I_q$ and I_M are determined: using estimates of the components of the flux linkage vector, as in previous models, and with using information about the position of the rotor plus the calculated value of the load angle δ. At that, in the first variant, the estimate is used when the rotational speed reaches at least 6% of the nominal one; no noticeable difference was observed. Since the estimation of the stator flux linkage at very low motor speeds, according to (1.17), (1.18), is performed with an error, in order to obtain a more or less accurate value of the rest tension, at a rotational speed of less than 6% of the nominal value, a unit value (in pu) of the excitation voltage is assigned, and only when the speed increases, the controller is switched to normal operation.

The torque reference, in addition to the tension reference and dynamic torque compensation when the speed changes, includes a number of other components. It is assumed that the torque of losses in the mechanism has a constant and a variable component: $M_{loss}\,(\text{pu}) = a + b\left|\omega_m^*\right|$. The value of the factor b is specified in the motor dialog box (it is taken $b = 0.02$), and $a = 0.04$. The torque spent for bending of the coiled strip under these conditions does not exceed 1% of the nominal and is neglected. One can also neglect the component of the dynamic torque which appears when the rotational speed of the coil changes with a change of the coil diameter at the constant reel-in speed.

Therefore, an accurate estimation of the existing diameter of the coil is of decisive importance. Usually, systems are used based on comparison of the signals of the strip speed V_m and the coil rotational speed ω_{dr}, since it should be $V_m = \omega_{dr}R_e$ where R_e is the estimation of the coil radius. The accuracy of such an estimate is determined by the accuracy of the measurement of the strip speed. When using the stand roll rotational speed signal, it is necessary to take into account the presence of the forward slip, which, in principle, can be done with a certain degree of accuracy using theoretical relationships or experimental data, since the tension values before and after the stand are known. To measure the speed of the strip, a measuring roller is often used, driven to rotation by the strip. But here one must keep in mind the possible slippage. Recently, laser meters for the speed of strip movement have gained ground [22].

In the model under consideration, a self-compensation method is used, in which the value of R_e is found by use of the relation

$$R_e = \int K_e (V_m - R_e \omega_{dr}) dt \tag{1.55}$$

where K_e is a gain. As it will be seen during simulation, the quantities R and R_e are practically equal. The block diagram of the assignment of the motor torque and calculation of the coil radius is shown in Figure 1.67.

A strip tension gauge T_m is installed often before the reel, but its use for the main tension controller causes difficulties due to the complexity of the controller structure and the variability of the parameters of the regulated object. It is preferable to use this meter to correct the I_q reference. Let's find the transfer function of the controlled object, taking the reference torque M*, referring to the drum shaft, as an input action. The controlled system equations are:

$$J \frac{d\Delta\omega_{dr}}{dt} = \Delta M^* - R\Delta T_m \tag{1.56}$$

$$\frac{d\Delta T_m}{dt} = \frac{Ebh}{L} (R\Delta\omega_{dr} - V_m \gamma \Delta T_m) \tag{1.57}$$

where $\gamma = \dfrac{\partial S}{\partial T_m}$ determines the change of the forward slip in the stand when the forward tension changes, J is the total moment of inertia of the rotating parts, reduced to the drum shaft, ω_{dr} is the speed of its rotation. After substitutions, it results as

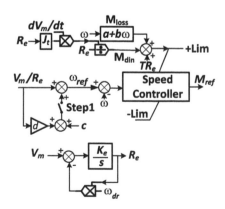

FIGURE 1.67 Block diagram of the assignment of the motor torque and calculation of the coil radius.

$$\Delta T_m = \frac{1/R}{J/AR^2 s^2 + J\gamma V_m/R^2 s + 1}\Delta M^* \tag{1.58}$$

where s is the Laplace operator symbol, $A = Ebh/L$. Let make an estimation of the transfer function parameters with $h = 1.6\,\text{mm}$, $b = 1610\,\text{mm}$, $L = 4.2\,\text{m}$ ($A = 1.22 \times 10^8\,\text{N/m}$), $V_m = 10\,\text{m/s}$, $\gamma = 0.03 \times 10^{-5}\,\text{1/N}$. When $R = 0.3\,\text{m}$, $J = 3700\,\text{kg-m}^2$,

$$\Delta T_m = \frac{3.33}{33.7 \times 10^{-5} s^2 + 0.123 s + 1}\Delta M^*,$$

damping factor $\xi = \dfrac{0.123}{2\sqrt{33.7 \times 10^{-5}}} = 3.35$; when $R = 1\,\text{m}$, $J = 23{,}266\,\text{kg-m}^2$,

$$\Delta T_m = \frac{3.33}{19.1 \times 10^{-5} s^2 + 0.07 s + 1}\Delta M^*,$$

$$\xi = \frac{0.07}{2\sqrt{19.1 \times 10^{-5}}} = 2.53.$$

Thus, with the accepted parameters of the reel, the controlled object is damped well, and a simple PI controller can be used to control the tension.

Models with indirect and direct tension control are investigated in the **Reel1** и **Reel2** models respectively.

The tension calculation by (1.31) for $h = 1.6\,\text{mm}$, $b = 1610\,\text{mm}$, $L = 4.2\,\text{m}$ is carried out in the subsystem **Reel**. To obtain stable operation when the rest tension is created, at zero and low speed, a value proportional to the input of the integrator is added to the output of the integrator (1.31) (otherwise, two integrators, (1.31) and (1.53), turn out to be connected in series in a closed system). The strip speed is computed as $V_m(1 + \gamma T_m)$, $\gamma = 3 \times 10^{-7}$, that is, at the maximal tension of 100 kN, the forward slip is 3%. The radius coil value is produced at the output of the integrator **Rad**: with each revolution of the coil, the radius increases by h mm, and to speed up the simulation, it is taken $h = 4\,\text{mm}$ at the input of the integrator (instead of the required value of 1.6 mm). This subsystem contains also circuits for the implementation of relation (1.54), as well as a coil radius meter according to (1.55).

In the model, the **Step1** block is activated at $t = 3\,\text{s}$, switching the speed controller to saturation mode, the value of the additional signal $c + d \times V_m/R_e$ is proportional to the actual speed reference. At the same time, the On signal is generated, activating the work of the tension and radius calculators. At that, the **Star Generator1** block sets the rest tension, and at $t = 4\,\text{s}$ it sets the working tension of the strip. Simultaneously, the strip speed quantity V_m begins to increase, at first, up to the speed of 0.5 m/s, with acceleration of 1 m/s², and afterwards, up to the speed of 10 m/s, with acceleration of 2 m/s². A **Rate Limiter**, with the opportunity to change the rate value by the external signal, is required, in order to realize such a program. First, the **Rate Limiter Dynamic** block from the Simulink library was used, but the simulation showed that a simulation speed is slowed down noticeably. Therefore, a special rate limiter is used in the following models, which was developed primarily for the mine electrical drives; its description is given in Figure 3.35.

The variable *Lim_pl* which determines speed controller limitation and, hence, the motor torque, is equal to the sum of the quantities: tension reference, dynamic torque compensation, the loss torque compensation, as it was described above (Figure 1.67).

The **Reel2** model differs from the **Reel1** model in that the output of the PI tension controller defines the limit *Lim_pl*, whereas in the **Reel1** model, this limit is defined by the tension reference and loss compensation signals; the dynamic torque compensation circuits are retained in the **Reel2** model, in order to facilitate the operation of the tension direct controller. In the **Reel1a** and **Reel2a** models, the DC sources for powering the inverter are replaced by active rectifiers with circuits that feed them, as it is the case in reality. Simulation in these models proceeds 1.5–2 times slower than in the original models, but it makes it possible to observe the currents, voltages, and powers of the network. The processes of the reel operation do not differ. It is worth noting that in the **Reel2a** model, the voltage in the DC link is reduced to 7500 V. Figure 1.68 shows the winding process (not completely) for the **Reel1** model, when, after acceleration to 10 m/s, the strip speed decreases to 8 m/s over a time interval of 12–27 s, and the tension decreases from 100 to 80 kN at $t > 27$ s. It can be seen that the tension value is quickly set after the start of speeding up and is subsequently constant both in static and dynamic conditions but with the error that is caused by unideal reproductions of the reference quantities. In principle, this error can be compensated with the proper correction of the tension reference value.

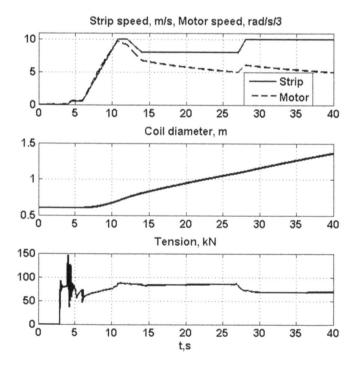

FIGURE 1.68 Process of winding in the **Reel1** model.

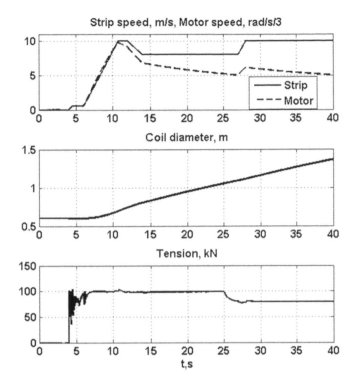

FIGURE 1.69   Process of winding in the **Reel2** model.

The same process for the **Reel2** model is shown in Figure 1.69. Tension control is carried out with greater accuracy; however, it must be borne in mind that the creation of an accurate and reliable tension meter is a rather difficult technical task. There are also systems in which both methods are used simultaneously, for example, the output of the tension regulator corrects the reference for the indirect regulator. In this case, it is possible to dispense with the circuits for compensation of the various kinds of losses.

The **Reel1** model provides the opportunity to simulate the process in the event of a strip break; the proper circuits are placed in the **Reel** subsystem. In this situation, under the action of the tension reference signal, the motor tends to increase the rotation speed having no counteraction from the load. With the adopted structure, when the rotational speed increases by the overtaking value, the speed controller leaves the saturation mode and the torque reference decreases. The process is shown in Figure 1.70 for two values of the coil diameter D, when, at $t = 15$ s for $R = D/2 = 0.3$ m and at $t = 20$ s for $R = D/2 = 0.8$ m, the strip breaks, and the torque of resistance decreases to the torque of loss. It can be seen that the rotational speed is effectively limited.

The reel operation as uncoiler is modeled in the **Reel3** model. Unlike the winding mode, when an increase in the rotational speed of the stand motor causes a decrease in tension, and an increase in the rotational speed of the reel motor causes tension increase, here the effect of these factors is opposite, which required a change in the signs of some signals. It is assumed that the thickness of the strip leaving the stand is 1.6 mm and the input thickness is 1.8 mm, which determines the speed of the strip

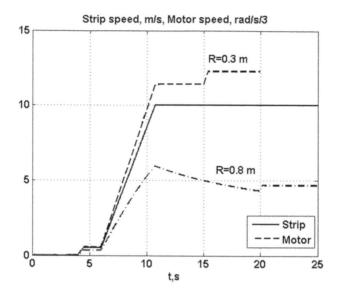

FIGURE 1.70   Process when the strip breaks.

on the uncoiler. When setting the tension, the additional signal is negative, and also the sign of the tension setting signal, which determines the negative speed controller limit level, changes. One fragment of the unwinding process is shown in Figure 1.71

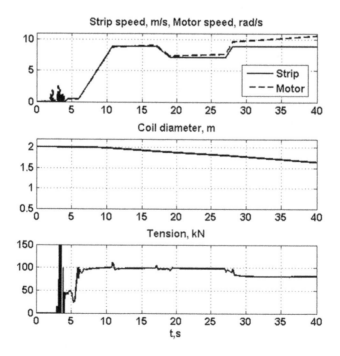

FIGURE 1.71   Fragment of the strip unwinding.

for the coil's initial diameter of 2 m. As for the **Reel1** model, after speeding up to 10 × (1.6/1.8) = 8.9 m/s, the strip speed reduces to 7.1 m/s over a time interval of 17–27 s, and the tension decreases from 100 to 80 kN at $t > 27$ s.

1.3.2 SIMULATION OF REVERSING COLD ROLLING MILLS

Reversing cold rolling mills are installed in cases when the volume of production is small and the range of rolled products is diverse. Most of these mills are single-stand, but recently two-stand reversing mills have also been built. Further, only single-stand mills are considered. Most of these mills have four-high stands, but when very thin strips are required, 12- and 20-roll stands are used. Only four-high mills are considered here, Figure 1.72.

Consider a mill with a working roll diameter $D = 2R = 600$ mm, a rolled strip width $b = 1610$ mm, and a rolling speed of up to 10 m/s. The stiffness of the stand is assumed to be 8 MN/mm. Rolling of a 6 mm strip to a thickness of 1.6 mm is carried out in five passes according to the schedule 6-4.38-3.11–2.4-1.8-1.6 mm. The specific tension is not less than 20 N/mm² (some decrease for unwinding in the first pass is allowed), and the total tension is not more than 160 kN.

Calculations of the rolling forces and torques are carried out at first. It has already been mentioned that when calculating the forces in the stands of a cold rolling mill, it is necessary to take into account the strain hardening of the metal during the rolling process; in contrast to hot rolling mills, it is possible to neglect the temperature and the deformation rate. The general relation for taking into account hardening can be represented as

$$\sigma_f = \sigma_s + a\varepsilon^b \tag{1.59}$$

where σ_s is a yield strength before deformation, a and b are the factors that depend on the rolled metal properties. These data are given in the various reference books (see Ref. [1]). So, it can be taken for the low-carbon steel $\sigma_s = 230$ N/mm², $a = 34.6$, $b = 0.6$, the value of ε in percent. Then, in accordance with the recommendations [2], since the degree of deformation changes from 0 to ε when the metal is reduced in rolls, the value $\varepsilon = 0.67\Delta h/h_0$ is used in (1.59). Thus, the formula (1.50) comes out.

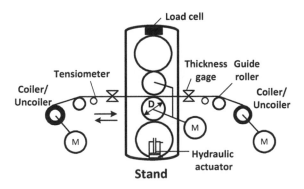

FIGURE 1.72   Design of four-high reversing cold rolling mill.

As already mentioned, there are various formulas for calculating the rolling force, and it is not clear which of them is more accurate in this case. Their use in a particular case is determined by the preferences and experience of the developer. A number of relations have already been given for the calculation of the roll forces; in cold rolling, the Stone method is quite popular [1]. In this method, the specific roll force

$$p_{pu} = 1.15\sigma_f \left(1 - \frac{\sigma_t}{1.15\sigma_f} \right) d \tag{1.60}$$

where σ_t is a half the sum of the back and forward specific tensions,

$$d = h_m [\exp\left(\frac{fl}{h_m} \right) - 1]/(fl), \tag{1.61}$$

$h_m = \dfrac{h_0 + h_1}{2}$, $\Delta h = h_0 - h_1$, $l = \sqrt{R\Delta h}$, f is the coefficient of friction,

The total roll force

$$P = p_{pu}bl \tag{1.62}$$

where b is the strip width.

However, these calculations do not take into account roll flattening, which is of great importance at high pressures during cold rolling. The amount of flattening depends on the rolling force, and this latter value in turn depends on the length of the contact of the rolls with the strip, taking into account flattening, so two equations must be solved together to determine the amount of flattening. Since we are mainly interested in electrical phenomena, we use a simplified method for finding a new contact area l_f.

$$l_f = x_1 + \sqrt{R\Delta h + x_1^2} \tag{1.63}$$

$$x_1 = 1.1 \times 10^{-5} \times R \times p_{pu} \tag{1.64}$$

where R is the undeformed radius (although the value of the deformed radius should be used in l_f for more accurate calculations). Then the rolling force

$$P_f = p_{puf}bl_f \tag{1.65}$$

where p_{puf} is computed by (1.60), (1.61), with using l_f instead of l.

The rolling torque is computed as

$$Tr = 2P_f y_f l_f. \tag{1.66}$$

Here

$$y_f = \psi \frac{p_{pu}}{p_{puf}} \frac{l}{l_f} \tag{1.67}$$

The value of ψ is calculated by (1.26)

For the computation of the rolling torque, the $Stone1$ program is created which is given for the second pass:

Program $Stone1$

```
H0=6;
h0=4.38;
h1=3.11;
R=600/2;
T0=160000;
T1=110000;
b=1610;
dH=h0-h1;
hm=(h0+h1)/2
r=dH/h0
l=sqrt(R*dH)
f=0.06
d=hm*(exp(f*l/hm)-1)/f/l
S0=230
S0=S0+34.6*(67*(H0-h1)/H0)^0.6
S=S0*1.15
K=d*S
sigma=(T0/b/h0+T1/b/h1)/2
Ppu=K*(1-sigma/S)
P=Ppu*l*b
x1=1.1e-5*R*Ppu;
lf=x1+sqrt(l^2+x1^2)
Rf=lf^2/dH
df=hm*(exp(f*lf/hm)-1)/f/lf
Ppuf=df*S*(1-sigma/S)
Pf=Ppuf*b*lf
m=f*l/hm
y=0.5*(1-r*(1+m)/(2+m))
%without roll flattening:
Tr1=2*P*y*l/1e6+R*(T0*h1/h0-T1)/1e6
yf=y*Ppu/Ppuf*l/lf
%with roll flattening:
Tr2=2*Pf*yf*lf/1e6+R*(T0*h1/h0-T1)/1e6
```

In the "**Appended models and programs**", this program is repeated for the last pass. Computation results are given in Table 1.8.

With such torque values, it is possible to use a motor operating with flux reducing and with an increase of the frequency of the converter in the last three passes. The synchronous motor is chosen with the data: 10 MW, 250 rpm, $Z_p = 10, f_n = 41.7\,\text{Hz}$, $J = 30{,}000$ kg-m^2,

$$H = \frac{3 \times 10^4 \times (2 \times \pi \times 41.7)^2}{2 \times 10^2 \times 10 \times 10^6} = 1.03\,\text{s},$$

TABLE 1.8

Rolling Schedule for the Reversing Cold Rolling Mill with the Initial Thickness of 6 mm

Pass	1	2	3	4	5
h_1, mm	4.38	3.11	2.4	1.8	1.6
T_0, kN	160	160	110	85	65
T_1, kN	160	110	85	65	56
P, MN	21	23.6	20.3	21.4	13.6
T_r, kN-m	373	379	250	235	94
N, rpm	250	250	315	315	315

motor drives the work rolls without gear. The motor nominal torque is

$$T_m = 10 \times 10^6 \times \frac{30}{\pi \times 250} = 382 \text{ kN-m}.$$ Therefore, the first two passes can be made at the rolling rated speed of $\frac{\pi \times 250}{30} \times 0.3 = 7.85$ m/s, perhaps, with some overload, taking into account the no-load loss. The rest of the passes are carried out with a frequency increase to 52.5 Hz at the rolling speed of 10 m/s.

The synchronous motors 3200 kW, 300 rpm, $Z_p = 6$, $M_m = 102$ kN-m, $f_n = 30$ Hz are chosen for the reels which drive the reel drums with reduction gears $i = 1.6$. The drum diameter is 600 mm, the coil maximal diameter is 2 m. Operating conditions of the reels in the modes winding/unwinding are given in Table 1.9. Recall that in order to distinguish between tension and torque for reel, the latter variable is denoted by M.

Let consider the first pass for the coiler, taking in mind formulae (1.51–1.53) and explanations to them. It is $T_b = 2 \times 102 \times 1.6/2 = 163$ кH (one takes $T_b = 160$ kN for simplicity); with an empty drum $\Psi* = 300/400 = 0.75$ and $D* = 0.3$, so that under $T* = 1$, $I_q^* = 0.4$(in fact, this value is greater by the amount of loss compensation). When the coil diameter increases, the rotational speed decreases, and the value of $\Psi*$ increases proportionally. With an increase of the diameter $D*$ to 0.4, the value of $\Psi*$ becomes equal to 1 and does not change when the coil diameter increases more. To maintain a given tension value, the value of I_q^* must increase in proportion to the increase in diameter.

TABLE 1.9

Operating Conditions of the Reels

Pass	V_Strip, m/s Uncoil/Coil	N_uncoil,rpm Full/Void	N_coil,rpm Void/Full	M_uncoil,kN-m Full/Void	M_coil,kN-m Void/Full
1	5.73/7.85	87.6/292	400/120	100/30	30/100
2	5.6/7.85	85.6/285	400/120	100/30	21/69
3	7.72/10	118/393	510/153	69/21	16/53
4	7.5/10	115/382	510/153	53/16	12/41
5	8.9/10	136/453	510/153	41/12	11/35

The moments of inertia of the motor and drum respectively: $J_{motor} = 2500$ kg-m^2 ($H_{motor} = 0.39$ s), $J_{dr} = 1200$ kg-m^2 ($H_{dr} = 0.19$ s). The moment of inertia of the coil reduced to the motor shaft is

$$J = \frac{\pi(D^4 - D_0^4)b \times 7800}{32 \times 1.6^2},\qquad\qquad (1.68)$$

and its maximal value is

$$J_{max} = \frac{\pi(2^4 - 0.6^4) \times 1.61 \times 7800}{32 \times 2.56} = 7643 \text{ kg-m}^2 (H = 1.18 \text{ s}).$$

Further, based on the analysis of the experimental results given in Ref. [23], the change of the forward slip with the change of tensions is taken in the form $\Delta S = g(\sigma_1 h_1 - \sigma_0 h_0)$, where σ_1, σ_0—are the forward and back specific tensions respectively, and $\Delta S = B(T_1 - T_0)$. If $g = 5 \times 10^{-4}$, then for the considered case $B = 30 \times 10^{-8}$.

The **Rev_Cold1** model simulates the stand drive in simplified way assuming that the reels on both sides of the stand provide ideal tension values; the same applies to the voltages in the DC link of the inverter. The model of the electric drive, its power supply, and control systems are the same as in the previously considered **Finish_Mill_Drive_10MW_300** model. The **Stand** subsystem calculates the forces and torques of rolling, as well as the output thickness for the given input thickness and tension values; it is made in the same way as it was done in the models of continuous hot rolling mills, with some differences related to strain hardening and flattening of the rolls, in accordance with the *Stone1* program. There are also circuits for adjusting the thickness of the rolled strip in this subsystem; these circuits use information of the rolling force sensor (with a possible error) and of the micrometer after the stand, and are similar to those used in hot rolling mills.

When the thickness controller is under investigation, it is assumed that the thickness variations at the entrance of the stand (in the first pass) can be modeled as a result of the passage of white noise through the element with the transfer function $W_f = (10s + 1)/(25s^2 + 5s + 1)$, which corresponds to the impulse response $k(t) = 0.4e^{-0.1t}\cos 0.17t$, implemented in a discrete form according to the formula of rectangles with a discrete period of T, specified in the *File/Model Properties/Callbacks/InitFcn* field.

Figure 1.73 shows the rolling process for the first pass. It can be seen that with sufficiently large deviations of the input thickness, the output thickness is maintained with a sufficient degree of accuracy, but this process is accompanied by significant fluctuations in the motor torque. This is probably not the case in practice.

Figure 1.74 shows the rolling process for the fifth pass, model **Rev_Cold1a**, also with an error in thickness estimation, because of a pressure sensor error of 5%. The thickness of the strip at the stand entrance changes stepwise at the time points of 15 s, 20 s, 25 s. Initially, the thickness of the strip at the exit differs from the specified one due to an error in the measurement of the rolling force, but after turning on the controller (at $t = 7$ s), it is equal to the specified value. Deviations in the output thickness caused by changes in the input thickness are quickly eliminated. It can be seen that the specified speed is achieved with a significant reduction of the motor flux.

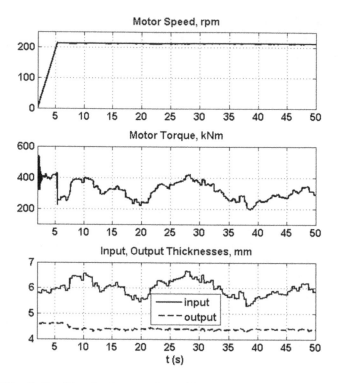

FIGURE 1.73 Cold rolling in the reversing mill with the large deviations of the input thickness.

The **Rev_Cold2** and **Rev_Cold2a** models model the rolling mill including a coiler and an uncoiler. The difference is that the first model assumes constant voltage sources in the DC link, while the second model contains an active input rectifier model and its power supply. Naturally, the simulation in it is much slower. The first model makes it possible to fairly adequately evaluate the processes in the electric drive and in the mill itself. However, simulation in it is also slow, so that only fragmentary results are given below.

Models of stand and reel drives have been considered above. The voltages of all three motors are assumed to be the same, equal to 3150 V. This allows the use of a common active rectifier for all three drives. The advantage of this approach is especially evident here, since both reels have approximately the same power, with one of them consuming power and the second giving power to the network, so that the total additional load from both reels for the input rectifier is small, and can even be negative. The reels with indirect tension control are used, though, of course, the variants with direct tension control can be employed as required.

The initial segment of the first pass is shown in Figure 1.75. At $t = 27$ s, the reference tension values are reduced by 20%. It can be seen that they are supported with an error of no more than 3%–4%. By changing the initial values of the coil radius, it is possible to obtain other fragments of the process; and by changing the given thicknesses of the rolled strip and, if necessary, the rolling speed, one can observe the processes in other passes.

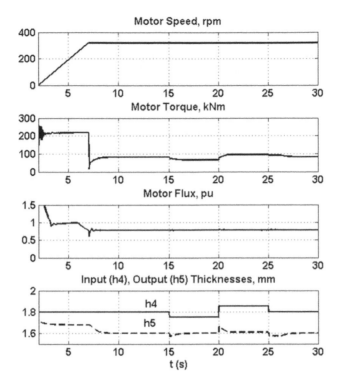

FIGURE 1.74 Cold rolling in the model **Rev_Cold1a.**

The fragment of the end of the running pass, when it is required to calculate the moment of the beginning of deceleration, taking into account the remainder strip on the uncoiler, is of certain interest.

The strip length at the stand input, during slowing-down with the deceleration of a_d, is $L_1 = h_1 V_m^2 / 2 a_d h_0$; the strip length that remains on the uncoiler and has n turns is

$$L_2 = 2\pi \left[nR_0 + h_0 n \frac{n-1}{2} \right] = 2\pi nR_0 \left[1 + \frac{h_0(n-1)}{2R_0} \right]. \tag{1.69}$$

Here h_0 and h_1 are the strip thicknesses at the stand input and output respectively, R_0 is the drum radius. For the first pass, $L_1 = 4.38 \times 7.85^2 / 6 \times 2 \times 2 = 11.25$ m ($a_d = 2$ m/s^2). If to take $n = 8$, then more than two turns ($L_2 = 16.12$ м) remain on the uncoiler after stop; that is sufficient for the mill reverse and to begin winding in the opposite direction. Thus, the command for slowing down must be given when the coil radius comes to $300 + 6 \times 8 = 348$ mm (subsystem **Stand/Control**).

To speed up simulation, the strip thickness is increased by six times, and only, when the coil radius at the uncoiler reduces to 0.4 m, the actual value of the thickness is restored. Figure 1.76 shows this process (model **Rev_Cold21**). It is seen that, under the complete stop, the coil radius on the uncoiler is 312 mm, that is, two turns remain, the rest tension is kept.

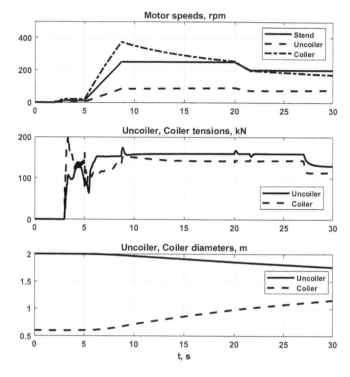

FIGURE 1.75   Beginning of the first pass in the complete model of the reversing cold rolling mill.

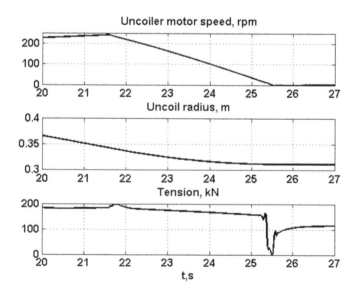

FIGURE 1.76   Stop process in the reversing cold rolling mill.

Note that this model, in order to demonstrate various approaches to modeling the system under consideration, has undergone significant changes compared to the **Rev_Cold2** model; namely, another method for modeling the changing moment of inertia of the load is applied. The method used in the **Rev_Cold2** model is simple but requires differentiation of the speed signal, which, if the model parameters are not chosen appropriately, can lead to an inaccurate estimate of the dynamic torque. In the method used in the **Rev_Cold21** model, the coiler drive is modeled as a two-mass system by applying the equations

$$\frac{d\omega_m}{dt} = \frac{M_m - M_{sh}}{J_{motor}} \tag{1.70}$$

$$\frac{d\omega_{dr}}{dt} = \frac{M_{sh} - M_{dr}}{J_{dr} + J} \tag{1.71}$$

$$M_{sh} = C\int (\omega_m - \omega_{dr})dt + B(\omega_m - \omega_{dr}) \tag{1.72}$$

Index m refers to the motor, dr refers to the reel drum, and C and B are coefficients characterizing the elastic properties of mechanical links. In our case, they are chosen arbitrarily, since we were only interested in taking into account the changing moment of inertia of the mechanism. Of course, if the exact values of these coefficients are known, and the consideration of the elastic properties of the mechanism is in the field of research interests, they can be used.

In the last pass, after the strip leaves the uncoiler, the rolls continue to rotate at low speed to release the end of the strip and thread a new one. Such a process is modeled in the **Rev_Cold22** model. Unlike the previous model, in this model, when the coil diameter on the uncoiler decreases to the definitive value, the given rolling speed is not equal to zero but is assumed to be 1 m/s. After the strip leaves the uncoiler, the tension is set to zero, and the rated motor excitation voltage is set. The length of the strip at the uncoiler during slowing-down is $L_1 = (10^2 - 1)/(1 \times 2 \times 1.6/1.8) = 44$ m. (Recall that the rolling speed in the last pass is 10 m/s, and the deceleration is assumed to be 1 m/s².) Thus, it can be taken that the speed decrease should begin when 22 turns leave on the uncoiler, that is, when, according to (1.69),

$$L_2 = 2\pi \times 22 \times 0.3 \times \left(1 + 1.8 \times \frac{21}{600}\right) = 44.06 \text{m},$$

that is, when the coil radius is $300 + 22 \times 1.8 = 340$ mm.

The corresponding process is shown in Figure 1.77 (as before, to speed up simulation, the strip thickness is increased by 20 times, and only, when the coil radius at the uncoiler reduces to 0.4 m, the actual value of the thickness is restored). It can be seen that at $t = 18.3$ s, the coil on the uncoiler decreases to a predetermined value, and the process of the mill slow down begins. At $t = 27.3$ s the speed of the strip reduces to 1 m/s, and the coil radius is 0.307 m. At $t = 35$ s, one turn of the strip remains on the uncoiler, the tension drops to zero, at $t = 38$ s it stops completely and is ready to

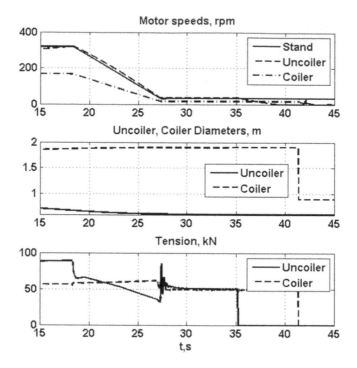

FIGURE 1.77 Stop process of the reversing cold rolling mill.

receive a new coil. The stand rolls carry on to rotate with the speed of 1 m/s. After a time interval equal to (4.2 + 1.9)/1 s (recall that the distance from the axis of the rolls to the coiler (uncoiler) is assumed to be 4.2 m and the length of one turn is 1.9 m), the strip leaves the stand, and the tension between the stand and the coiler drops to zero. The coiler continues to rotate for about the same time to fully wind the coil, after which it stops.

As already mentioned, the **Rev_Cold2a** model (as well as **Rev_Cold2a1**), in addition to the previous models, contains an active rectifier with its power supply system, which forms 7000 V DC link. The rectifier control system is taken from the previous models; it maintains the wanted value of the DC voltage by adjusting the current drawn from the network. All three inverters are connected in parallel to this voltage, Figure 1.78. Note that, in order to speed up the simulation, only one three-level converter is installed in the active rectifier model, while for the required power it is necessary to have at least two converters, as is done to power the stand motor. The parameters of the first model correspond to the first pass, and the second to the fourth. A start of the first pass is shown in Figure 1.79; the reactive power controller is on. It is supposed that the rolled strip thickness is estimated with an error and is adjusted with the help of the micrometer after the stand. It is seen that the active power consumed from the network, at the end of speeding up, reaches 20 MW, the reactive power is close to null, and THD in the network current is about 4.5% after the acceleration. About the same results occur for the fourth pass.

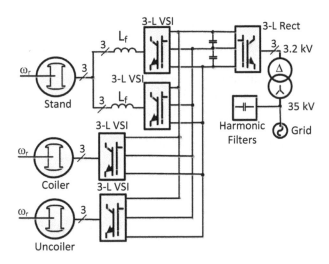

FIGURE 1.78 Diagram of circuits for fabrication of the DC link voltage.

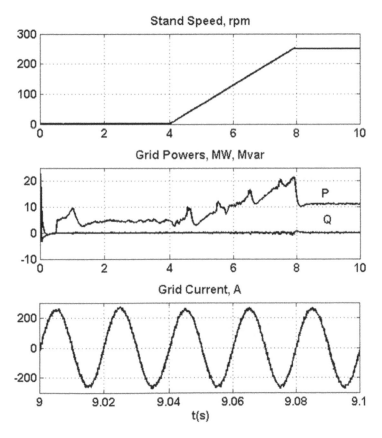

FIGURE 1.79 Start of the first pass in the reversing cold rolling mill.

1.3.3 SIMULATION OF THE CONTINUOUS COLD ROLLING MILL

In the end of the chapter, modeling of continuous cold rolling mills—Tandem mills, which produce most of the cold rolled sheets, is considered. Mills of this type can have four or five stands (the former are considered obsolete). The general arrangement of the five-stand mill is shown in Figure 1.80. It is assumed that tension gauges are installed in the inter-stand space, and strip thickness gauges are installed after the first and fifth stands.

The five-stand mill with a diameter of the work rolls $D = 616$ mm will be modeled, the coil mass is 45 t. The individual drive of the stand rolls is used, with the same motors for all stands and with different gears, Figure 1.81.

An employment of the synchronous motors 4500 kW, 600 rpm ($Z_p = 5$), 71.65 kN-m is considered. Let take $i_1 = 1.75$, $i_2 = 1.75$, $i_3 = 1.25$ $i_4 = 0.91$ $i_5 = 0.91$. Two rolling schedules, which were realized on the real rolling mill, are considered [24]. The tables below show the thicknesses h_1 (mm) at the exit of the stands, the reference tensions after the stands T_1 (kN), the rolling speeds in the stands V (m/s) and the roll speeds N_r (rpm), the rolling forces P (MN) and torques M (kN-m), calculated using the program Stone1, as well as the values of the speeds of rotation of the motors N_m, their nominal torques at this speed of rotation M_m and the values of these torques, reduced to the rolls M_r (obviously, it should be $M_r > M$ with some margin). Friction coefficient $f = 0.065$.

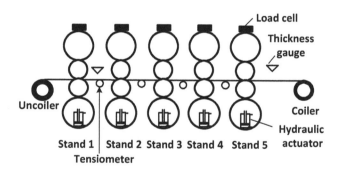

FIGURE 1.80 General arrangement of the five-stand cold rolling mill.

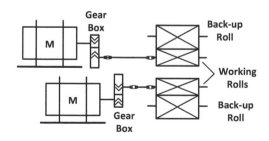

FIGURE 1.81 Arrangement of the stand drive of the continuous cold rolling mill.

One separate stand is modeled in the **Tand1** model. The model of the electric drive of the rolls with its control system is taken from the **Rev_Cold1** model with some minor changes caused by the need for several such models to function together when modeling the mill as a whole. The parameters corresponding to the rolling schedule and stand number are set in the *Model Properties/Callbacks/InitFcn* window. The applied **Step** blocks make it possible to investigate the change in parameters, in particular, the change of the thickness at the exit of the stand when the tension or input thickness changes. It is possible to investigate the thickness controllers using the measurement of the rolling force, with and without correction by the micrometer signal, as well as with a constant setting of the roll clearance S_0.

As an example, Figure 1.82 shows the initial segment of the fifth stand operation when rolling the second range (0.7 × 1100 mm). At $t = 18$ s, the back tension of the fifth stand is increased by 20%, that is, by 22 kN. The output thickness decreases by 5 µm, that is, ratio $\Delta h_5/\Delta T_{4-5} \cong 0.23$ µm/kN, which is the same order as the experimental data given in Ref. [23]. Changes of the active and reactive powers and the network current for the process recorded in the previous figure are shown in Figure 1.83; the reactive power controller is turned on. It can be seen that the reactive power is close to zero, the THD of the network current does not exceed 3% with the voltage in the DC link of 2 × 3.5 kV.

When modeling the entire mill, it must be borne in mind that it proceeds extremely slowly, therefore, in a number of the following models, the stand drive is modeled in a simplified way, as has already been applied in modeling continuous

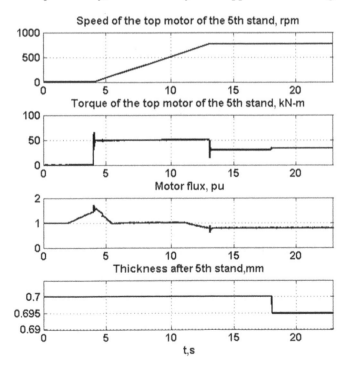

FIGURE 1.82 The initial segment of the fifth stand operation when rolling 0.7 × 1100 mm strip.

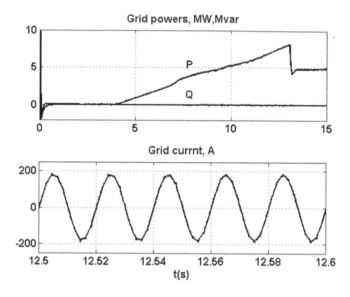

FIGURE 1.83 The grid powers and current when rolling $0.7 \times 1100\,\text{mm}$ strip.

hot rolling mills: as a serial connection of a first-order element, the time constant of which is determined by the drive response time to change the torque reference (it is the output of the speed controller), and the equation of torques (1.11). Simulation of the separate stand (**Tand1** model) proves that the time constant of the first order element can be taken equal to 8 ms.

In the subsequent simulation for the rolling of $0.7 \times 1100\,\text{mm}$, the value of the yield strength before rolling was taken equal to $230\,\text{N/mm}^2$, as for the rolling of $1 \times 1850\,\text{mm}$. In this case, the schedule given in Table 1.11 is replaced by the schedule in Table 1.12.

The **Tand2ab** model simulates a five-stand mill, the parameters of which are given above. In the stand model dialog box, the nominal output strip thickness, the gear ratio of the motor to the work rolls, and the nominal value of the rolling force used to calculate the position of the pressure screws are indicated; the rolling force is calculated using the *Stone1* program. Other stand parameters are given in the dialog box of **Subsystem** included in the **Stand** subsystem, which has already been used many times in previous models. The **Drive** subsystem includes a speed controller

TABLE 1.10

Rolling Schedule 1×1850 for $h_0 = 3.3\,\text{mm}$, $\sigma_s = 230\,\text{N/mm}^2$

h_1 (mm)	T_1 (kN)	V(m/s)/ N_r(rpm)	P(MN)/M (kN-m)	N_m (rpm)	M_m(kN-m)/ M_r(kN-m)
2.5	310	4.84/150.1	17.7/138	262.7	143.3/250.7
1.85	270	6.54/202.7	19.3/211.1	354.7	143.3/250.7
1.43	240	8,46/262.2	18.2/160.4	327.8	143.3/179.1
1.15	240	10.52/326.2	16.6/111.8	296.8	143.3/130.4
1.0	90	12.10/375.2	14.2/117.5	341.4	143.3/130.4

TABLE 1.11

Rolling Schedule 0.7 × 1100 for h_0 = 2.6 mm, σ_s = 300 N/mm²

h_1 (mm)	T_1 (kN)	V(m/s)/ N_r(rpm)	P(MN)/M (kN-m)	N_m (rpm)	M_m (kN-m)/ M_r(kN-m)
2.03	230	9.38/290.8	10/65.7	508.9	143.3/250.7
1.39	190	13.7/424.8	14/140.1	743.4	115.6/202.4
1.01	150	18.85/584.5	13.2/105.3	730.6	117.7/147.1
0.79	110	24.10/747.3	12/78.1	680	126.4/115.1
0.7	60	27.20/843.3	9.5/50.9	767.4	112.1/102

taken from previous models and a simplified electric drive model. Inter-stand models include a delay block simulating the passage of changes in strip thickness between stands and a tension calculator, the same as used in models of hot rolling mills. The dependence of the forward slip on tensions is assumed to be the same for all stands, with the parameters given above. The distance between the stands is assumed to be 4.2 m. In the *Model Properties/Callbacks/InitFcn* field, parameters common for the entire mill are given: strip width *b*, specified exit strip thickness h_{05}, yield strength before rolling *Sig*, and strip thickness at the mill entry H_0. It is assumed that the tensions of the coiler and uncoiler are maintained with high precision by their control systems. The model provides for the possibility of simulating interstand tension controllers that affect the rotational speeds of the previous stands, as well as the regulation of the strip thickness at the mill exit.

The **Tand2ab** model simulates the rolling of 0.7 × 1100 strip, the rolling schedule is shown in Table 1.12. In this model, the speeds of rotation of the motors and the roll clearances are chosen in such a way that the values of thicknesses and tensions are equal to the given values without the participation of regulators. This model can be used to determine the process gains of the mill, which are the ratio of changes in controlled values to the changes in control actions that caused them. As the former, changes in the thickness of the strip after the stands and inter-stand tensions are considered, and as the second, changes in the stand motor rotational speeds and changes in the position of the work rolls. Knowledge of these values is used in the development of mill control systems. Comparison of the simulation results with the

TABLE 1.12

Rolling Schedule 0.7 × 1100 for h_0 = 2.6 мм, σ_s = 230 Н/мм²

h_1 (mm)	T_1 (kN)	V (m/s)/ N_r(rpm)	P (MN)/M (kN-m)	N_m (rpm)	M_m (kN-m)/ M_r(kN-m)
2.03	230	9.38/290.8	8.4/30.7	508.9	143.3/250.7
1.39	190	13.7/424.8	11.5/114.7	743.4	115.6/202.4
1.01	150	18.85/584.5	10.9/87.2	730.6	117.7/147.1
0.79	110	24.10/747.3	9.92/65.4	680	126.4/115.1
0.7	60	27.20/843.3	7.65/44.1	767.4	112.1/102

experimental data given, for example, in Refs. [23,25], makes it possible to judge the adequacy of the developed model.

The reference [25] provides data on two four-stand cold rolling mills: 1680 and 1700 in the USSR. So, for example, it is indicated that when the roll clearance of the first stand dS_1 reduces all interstand tensions increase and the thickness of the strip at the exit of the mill decreases. If the simulation is performed with a decrease of dS_1 by 15 μm at $t = 12$ s, then $dh_1 = 5$ μm, $dh_5 = 1.5$ μm, $dT_{1-2} = 800$ N, that is, $dh_1/dS_1 = 0.33$, $dh_5/dS_1 = 0.13$, $dT_{1-2}/dS_1 = -53$ kN/mm. Average readings from Ref. [25]: $dh_4/dS_1 = 0.15$, $dT_{1-2}/dS_1 = -110$ kN/mm, but it is a completely different mill. It is ascertain in Ref. [23] that, with roll clearance decrease of the rest stands, the back tensions of these stands reduce very much but the other tensions change much less. Simulation confirms that assertion. Thus, for example, when the roll clearance of the second stand reduces by 0.047 mm, the tension T_{1-2} decreases by 9 kN, so that $dT_{1-2}/dS_2 = 191$ kN/mm, when the roll clearance of the third stand reduces by 0.06 mm, the tension T_{2-3} decreases by 12 kN, so that $dT_{2-3}/dS_3 = 200$ kN/mm, when the roll clearance of the fourth stand reduces by 0.061 mm, the tension T_{3-4} decreases by 10 kN, so that $dT_{3-4}/dS_4 = 164$ kN/mm, and when the roll clearance of the fifth stand reduces by 0.0437 mm, the tension T_{4-5} decreases by 5.8 kN, so that $dT_{4-5}/dS_5 = 133$ kN/mm. The order of quantities is the same in Ref. [23]. It is also noted there that the factor dh/dS in the last stand is small. Indeed, modeling shows that in this case $dh_5/dS_5 = 0.046$. Then, the simulation shows that increase of the rotation speed of the fifth stand by 1% at a rolling speed of 27.2 m/s increases the tension by 8 kN and reduces the exit thickness by 5 μm. When performing these experiments, the tension and thickness controllers were turned off (the **Manual Switch** tumblers in the **Cold Rolling Stand** models and the **Switch1** tumblers in the **Stand** subsystems were in the lower positions, the **Manual Switch** tumbler on the model diagram in the upper position, the **hin1** block sets the input thickness of 2.6 mm).

This model provides the opportunity to simulate the eccentricity of the backup rolls of the first stand, with use of the same technique as previously used for continuous hot rolling mills. To do this, in the model of the first stand, the **Stand** subsystem, the **Switch2** tumbler is set to the lower position, the eccentricity amplitude is set in the **Ecc** subsystem, and the oscillation frequency is calculated in the **Gain2** block. The diameter of the back-up rolls is assumed to be $D_b = 1.6$ m. The oscillation frequency $f = V_m h_5 / (h_1 D_b \pi)$. With an amplitude of 50 μm, the amplitude of thickness fluctuations after the first stand is 10 μm, and after the fifth stand is 5 μm. The amplitudes of fluctuations in rolling forces and motor torques are small, but, nevertheless, with their frequent repetitions, the wear of mechanical parts can be intensified, especially when the perturbation frequency coincides with the natural frequencies of the mechanical system.

In cold rolling mills, the so-called speed effect is observed, which consists in a decrease in the thickness of the rolled strip with an increase in rolling speed. The main reasons for the speed effect are a decrease in the thickness of the oil film in the fluid friction bearings of the rolls and a decrease in the coefficient of friction. According to the data given in Ref. [1], this decrease can be quite significant, almost twice. The **Tand2ab1** model is intended for a preliminary study of this phenomenon. Compared to the previous model, some changes have been made to it. In particular, the friction coefficient f is not assigned in the dialog box of the stand model but is determined, depending on the rolling speed, in the subsystems **f_factor** in the models **Cold rolling Stand**. There is also an input for changing the roll clearance as a function of speed. When modeling, it is assumed that with an increase in the rolling speed from zero to 27.2 m/s, the roll clearance will decrease by 20 μm. The factor f decreases from 0.65 to 0.43 with an increase in speed up to 15 m/s and does not change with a further increase in speed.

Changes in some thicknesses and tensions are shown in Figure 1.84. It can be seen that all these variables decrease with the speed increase, which is consistent with the experimental data given in Ref. [23]. It is interesting to note that, in percentage ratio, the change in tension is significantly greater than the change in thickness.

Mill control systems have to maintain the specified strip thickness at the mill output and restrict changes of interstand tensions by specified limits. Figure 1.85 shows the process of passage through the stands of the mill perturbation of the input thickness by 5% at $t = 12$ s. Thickness and tension control systems are disabled. It can be seen that thickness deviations in all stands begin almost at the same time, since all interstand tensions begin to change immediately. It is interesting that after the end of the transient processes, the tension returns to the initial values.

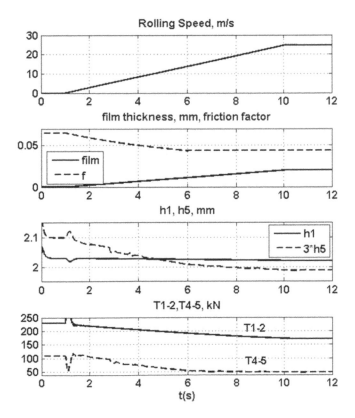

FIGURE 1.84 Manifestation of the speed effect.

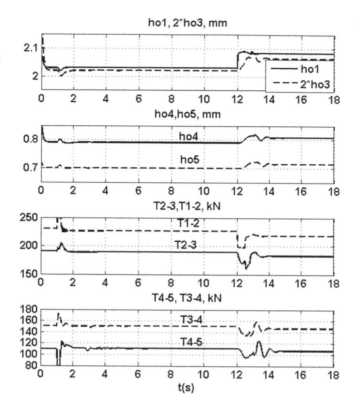

FIGURE 1.85 Process of passage through the stands of the mill perturbation of the input thickness without controllers.

System various structures for controlling thicknesses and tensions are described in the technical publications; they differ in the types and number of measuring parameters of the rolling process and the direction of the control actions. A number of such systems are described in Refs. [22,23,26–28]. As a new trend, one can note the use of laser strip linear velocity meters in various interstand spaces, which allows, using the principle of metal flow continuity, to obtain the value of the strip thickness in each interstand space without delay and without the use of a micrometer [22,27]. The aim here is not to simulate a complete control system but to demonstrate the basic principles.

It has already been said when considering continuous hot rolling mills that continuous mills are a united electromechanical complex, where all process parameters are interrelated. Therefore, it seems logical to analyze such a mill as a unified system described by a system of high-order equations and find relationships that optimize such a system. This approach was applied in Ref. [28], where a five-stand mill is described by a system of equations of the 14th order. The disadvantages of this approach have already been discussed. It can be noted that this method is based on the relations describing the rolling process; there are various relationships of this type, and with the utilization of other relationships, the optimization result may also be different; it remains unclear what is the equivalence between the optimization results and the actual process. Therefore, in practice, methods for the development of the control systems based on ideas about the physical processes in the system are used.

First, systems for control of the interstand tensions, with an action on the rotational speed of the previous stand, are modeled. The change of the tension between stands 1 and 2 can be written as

$$\frac{dT_{12}}{dt} = \alpha \left[V_2 \frac{h_2}{h_1}(1 + S_0 + \gamma(T_2 - T_{12})) - V_1(1 + S_0 + \gamma(T_{12} - T_1)) \right]. \quad (1.73)$$

We take $S_0 = 0.04$, coefficient γ as $\gamma = 5 \times 10^{-4}/b$ N^{-1}, b is a width, mm, for $b = 1100$ mm $\gamma = 45.5 \times 10^{-8}$, $\alpha = \dfrac{Ebh_1}{L}$, E is a modulus of elasticity, L is a space between stands. The back tension for the first stand T_1 and the forward tension for the second stand T_2 are taken as the given quantities.

It is in increments, supposing the variables T_1, T_2, h_1, h_2, V_2 constant:

$$\frac{1}{\alpha}\frac{d\Delta T_{12}}{dt} = -(D_1 + D_2)\Delta T_{12} - \Delta V_1 S_{10} \quad (1.74)$$

$$D_1 = \gamma V_{10}, \quad D_2 = \gamma V_{20}h_2/h_1, \quad S_{10} = 1 + S_0 + \gamma(T_{12\_0} - T_1).$$

If to take $V_{20} h_2/h_1 \approx V_{10}$, it can be written finally

$$T_l \frac{d\Delta T_{12}}{dt} + \Delta T_{12} = -K_l \Delta V_1 \quad (1.75)$$

$$T_l = \frac{L}{2Ebh_1\gamma V_{10}}, \quad K_l = \frac{S_{10}}{2\gamma V_{10}}.$$

$$\text{Or } T_l \frac{d\Delta T_{12}}{dt} + \Delta T_{12} = -K_{l1}\Delta V_1^* \quad (1.76)$$

$K_{l1} = S_{10}/2\gamma$, ΔV_1^* is relative change in the speed of the first stand.

Let's take parameters of stands 4 and 5 with $V_2 = 27.2$ m/s, $h_1 = 0.79$ mm, $h_2 = 0.7$ mm, $T_1 = 150$ kN, $T_{12\_0} = 110$ kN, $T_2 = 60$ kN, $b = 1100$ mm, $S_{10} = 1 + 0.04 + 45.5 \times 10^{-8}$ $(110,000 - 150,000) = 1.02$, $K_{l1} = 0.5 \times \dfrac{10^8 1.02}{45.5} = 1.12 \times 10^6$, so that under speed change by 1% the tension change will be 11.2 kN = 10%. The time constant under $V_{10} = 27.2 \times 0.7/0.79 = 24.1$ m/s is equal to

$$T_l = \frac{4.2}{2 \times 0.2 \times 10^6 \times 1100 \times 0.79 \times 45.5 \times 10^{-8} \times 24.1} = 1.1 \text{ ms.}$$

If to carry out the simulation, assuming that the input and output tensions and thicknesses of stands 4 and 5 do not change, then, with a decrease in the speed of stand 4 by 1%, the increase in tension will be ~12 kN, that is, close enough to the calculated value. Of course, when considering the entire mill as a whole, the value of the tension change will differ due to changes in tensions between other stands and strip thicknesses. This value was found to be only 6 kN.

The tension controllers which act on the rotational speeds of the motors of the previous stands are simulated in the **Tand2ab** model. Tension controllers are PI type with limitations and are activated when the rolling speed is equal to approximately 10% of the maximum. The rate of the action is proportional to the current rotational speed of the motors.

The same process as in Figure 1.85 is shown in Figure 1.86 but the tension controllers are turned on (the **Manual Switch** tumblers in the **Cold Rolling Stand** models are in the up position). It can be seen that the tension changes decrease. In particular, the maximum reduction in tension T_{1-2} has changed from 30 to 17 kN, and tension T_{2-3} from 28 to 10 kN. Figure 1.87 shows the process when the tension between stands 1 and 2 changes by 20%. It can be seen that the other tensions, after small short-term deviations, return to the specified values.

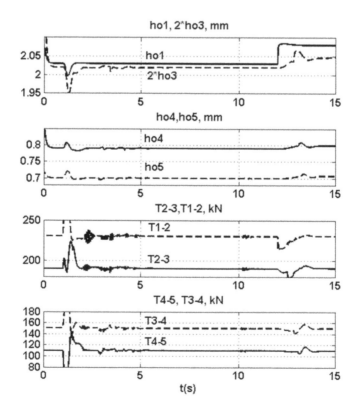

FIGURE 1.86 Process of passage through the stands of the mill perturbation of the input thickness with turned-on tension controllers.

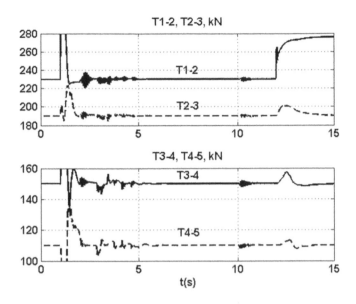

FIGURE 1.87 Process when the tension between stands 1 and 2 changes.

The tension controllers that actuate on the roll clearance of the next stand are simulated in the **Tand2ab2** model. The tension controllers that actuate on the rotational speeds of the motors of the previous stands are turned off. The PI controllers are placed in the **Stand** subsystems of models of stands 2–5. The controller outputs change the reference values of the roll clearances at the inputs of the hydraulic drive models. The same process as in Figure 1.86 is shown in Figure 1.88. It can be seen that the tension changes are almost the same as in Figure 1.86: the maximum decrease in tension T_{1-2} is 20 kN, and tension T_{2-3} is 11 kN. The process is fixed in Figure 1.89 when the tension setting T_{1-2} is changed by 20%, as in Figure 1.87. It can be seen that the type of change of the other interstand tensions differs from those shown in 1.87.

The main indicator of the quality of the strip cold rolling is its longitudinal variation in thickness. Therefore, the creation of perfect thickness controllers is of great importance. As can be seen from the above given values of the process gains, the effect on the rotational speed of the rolls and the position of the rolls of the last stands of the cold rolling mill is not effective, which is determined by a significant amount of strain hardening of the metal. Therefore, when developing a control system, the main attention is paid to reducing the thickness variation caused by deviations in the thickness of the hot-rolled strip entering the cold rolling mill. The actuator of this control system is mainly the hydraulic drive of the displacement of the rolls of the first stand. There are different structures of such systems: with one micrometer behind the first stand, with such a micrometer and a rolling force gauge in the first stand, with two micrometers: behind the first and second stands, with two micrometers: before and after the first stand, and so on. Subsequent models use a control system based on measuring the rolling force in all stands (possibly with an error), with correction based on indications of the micrometers behind the first and fifth stands. To obtain such a structure of the control system, the **Switch1** tumblers in the **Stand**

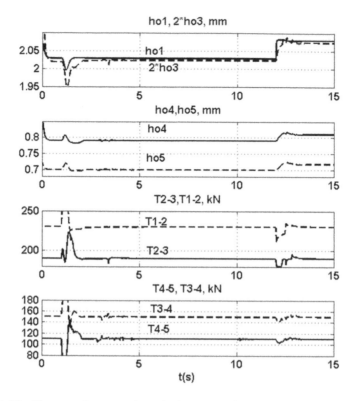

FIGURE 1.88 Process of passage through the stands of the mill perturbation of the input thickness, with tension controllers acting on the roll clearances.

subsystems of the stands models are set to the upper position, the same applies to the **Switch** tumblers in the same subsystems of the first and fifth stands. The tension controllers are turned on.

When simulating thickness control in the **Tand2ab** model, it is assumed that the exit thickness measurement from the rolling force measurement is performed with an error of 5%. There are various options for specifying input thickness deviations. The process is shown in Figure 1.90 when these deviations change from −0.05 to 0.13 mm with a period of ~6 s. At $t = 22$ s, during the passage of the weld, the thickness increases by 0.4 mm with a duration of 2 s. To do this, the **Manual Switch** tumbler on the model diagram is set to the lower position, and the **Manual Switch1** tumbler to the left position. It can be seen that noticeable thickness deviations after the first and fifth stands occur only before turning on the regulators with micrometers, while after the fourth stand the thickness noticeably differs from the specified one (0.79 mm) due to inaccurate measurement of the rolling force. During the passage of the weld, small deviations occur that are quickly parried by the control system.

However, this model does not take into account the presence of the speed effect. Its manifestation is demonstrated in the previously considered **Tand2ab1** model. In the **Tand2ab4** model, the speed effect is simulated in the presence of all previously considered controllers: control of the strip thickness according to measurements of

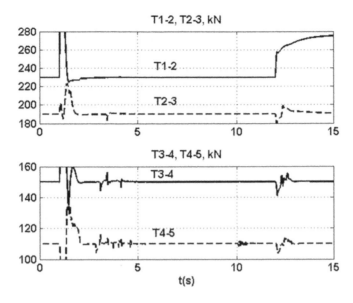

FIGURE 1.89 Process when the tension between stands 1 and 2 changes, with tension controllers acting on the roll clearances.

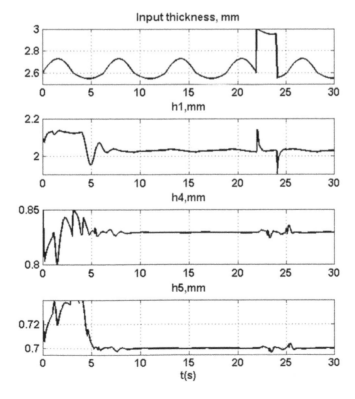

FIGURE 1.90 Process of the thickness control on the continuous cold rolling mill.

the rolling forces with an error of 5%, corrections based on micrometer indications behind the first and fifth stands, tension controllers that affect the rotational speed of the motors of the previous stands.

The process is shown in Figure 1.91. It can be seen that despite the errors in measuring the rolling force, after turning on the micrometers, the strip thickness after the first and fifth stands is equal to the specified values, and the speed effect is compensated. In this case, the rolling speed changes, as shown in the upper plot of Figure 1.84.

The plots in Figure 1.92 demonstrate the possibility of controlling the output thickness by the action on the rotational speed of the motors of the last stand. The **Tand2ab5** model is used. At this, the tension controller, which affects the rotational speed of the motors of the fourth stand, is turned off. The block diagram of the controller that affects the rotational speed of the fifth stand is shown in Figure 1.93. The PI tension controller, which in previous models affected the speed of the fourth stand motor in the deceleration direction (when it was necessary to increase tension), now affects the speed of the fifth stand motor in the opposite direction. The tension setting is corrected by the output of the thickness controller; for a feedback, the signal of the micrometer located at a distance of 1.5 m after the fifth stand is used. The employment of an integral block as a thickness controller in this case is impractical, since in order to return it to its initial state, it is necessary to change the sign of the thickness error, which in this case may not happen. Therefore, a first-order element

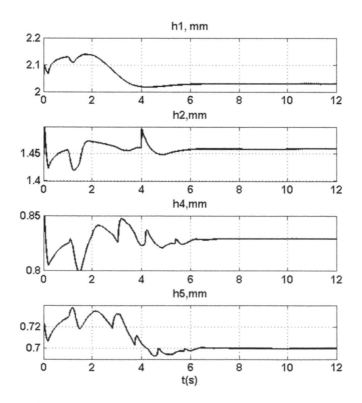

FIGURE 1.91 Thickness control with taking into account the speed effect.

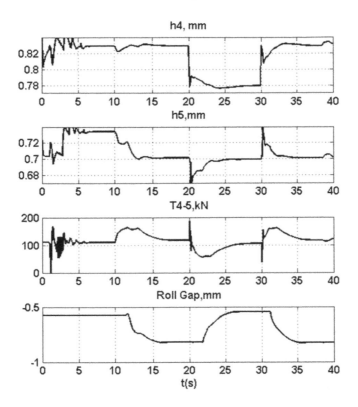

FIGURE 1.92 Thickness control by the action on the last stand motor rotation speed.

with a time constant of 1 s is used as a controller. When the deviation of the tension value from the specified one reaches 35% of the last value, the *H_On* signal is generated, which turns on the thickness controller; this controller affects the roll clearance and was used in previous models. When the tension deviation decreases to 10%, the *H_On* signal is removed and the roll clearance is fixed. Figure 1.92 shows the process when the thickness controller is turned on at $t = 10$ s. It can be seen that before the controller is turned on, the output thickness differs significantly from the specified one due to an error in the stand stiffness estimation. After turning on the controller, the rotational speed of the motor of the fifth stand increases, which leads to an increase in the tension between the stands, the thickness deviations decrease, but nevertheless the specified thickness is not reached, so the thickness controller is turned on which affects the roll clearance. The output thickness becomes equal to the wanted value and the magnitude of tension deviation from the set decreases. At $t = 20$ s and $t = 30$ s, significant abrupt changes in the input thickness take place, which initially leads to noticeable changes in the interstand tension, but then the controller comes into action, changing the roll clearance, and thickness deviations are reduced to zero.

The **Tand3** model simulates the rolling of strip 1 × 1850 mm, the schedule of which is given above in Table 1.10. Figure 1.94 shows the process of passage through the mill of a disturbance of the input thickness of 5% with the thickness controllers turned off, as in Figure 1.86 for the schedule 0.7 × 1000 mm.

If it is required to use complete models of stands drives, then it is necessary to replace in the described models the simplified models of stands with detailed ones, for example, the one used in the **Tand1** model; however, simulation will proceed

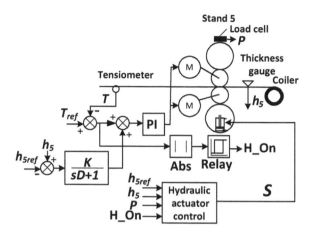

FIGURE 1.93   Block diagram of the controller that affects the rotational speed of the fifth stand.

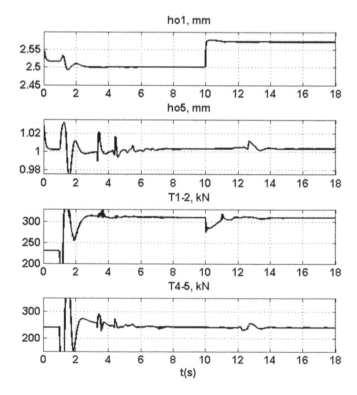

FIGURE 1.94   Process of passage through the mill perturbation of the input thickness with turned-on tension controllers for the rolling of strip 1 × 1850 mm.

extremely slowly in such a model. To estimate the impact of the mill electric drive on the supply network, the **Tand4b** model can be used. In this model, all ten stand motors are replaced by one equivalent motor with a total power of 49 MW (including the coiler) with the same relative parameters. The motor connects to the 35 kV network with the help of the active rectifier with three-level VSI and the transformer with voltage of 35/3.5 kV and power of 75 MVA. The motor and inverter control systems were described above. The load power of the equivalent motor is equal to the total load power of all mill motors.

In order to determine this power as a function of time, the mill simulation is carried out as it was made in the **Tand2ab** model, with an additional record of the load powers which are the products of the stand motor load torques by their current rotational speeds. These augmented models are simulated in the **Tand4** model for the rolling schedule 0.7 × 1100 mm and in the **Tand41** model for the rolling schedule 1 × 1850 mm. The coiler power $T_{coil} \times V_m$ is added to the stand total power where the first multiplier is the coiler tension. When simulation is carried out, the resulting power is fixed in the matrix **SC20**, and the rolling speed in the matrix **Vm**. After simulation, its result is stored in the file $Tand\_Load1$ for the first schedule and in $Tand\_Load2$ for the second.

When the first rolling schedule is simulating, the instruction $load('Tand\_Load1')$ is given in the field *Modeling/Model Settings/Model Properties/Callbacks/*

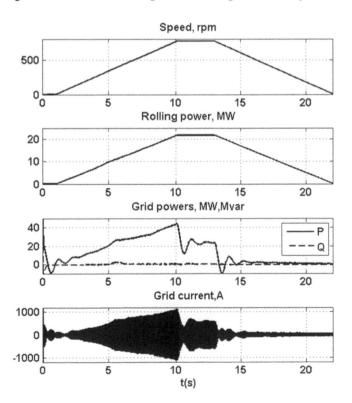

FIGURE 1.95 Rolling 0.7 × 1100 mm with equivalent motor.

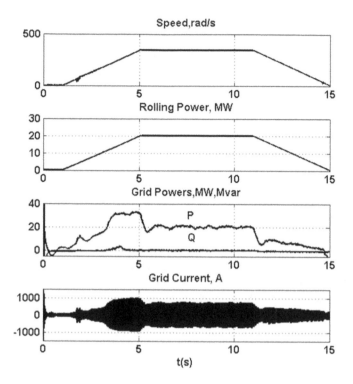

FIGURE 1.96   Rolling 1 × 1850 mm with equivalent motor.

InitFcn of the **Tand4b** model, and the instruction `load('Tand _ Load2')` when simulating the second schedule. The **From Workspace** block addresses are given as *SC*20 и *Vm*. The simulation time is set as 22 s in the first case and 15 s in the second one. The simulation result for the rolling 0.7 × 1100 mm with the network reactive power controller turned on is shown in Figure 1.95. The processes of acceleration, steady-state rolling speed and deceleration are shown; the duration of the steady speed is much less than the actual one, but if necessary it can be easily increased. It can be seen that the reactive power is close to zero, and the value of the active power consumed from the network reaches 42 MW at the end of acceleration. The THD value of the mains current at this instant of time is less than 3%. A detailed analysis of the network current waveforms allows selecting the structure and parameters of network filters, as well as the reasonability of using other devices to improve the quality of electricity: active filters, STATCOM devices, and so on.

The result of simulation of rolling 1 × 1850 mm is shown in Figure 1.96.

REFERENCES

1. Zyuzin, V. I., Tretyakov, A. V., ed. Rolling theory (in Russian). Handbook. Moscow, Metallurgy, 1982.
2. Polukhin, P. I., Fedosov, N. M., Korolev, A. A., Matveev, Y. M. Rolling Production (in Russian). Metallurgy, Moscow, 1968.

3. Benyakovsky, M. A., Bogoyavlensky, K. N. etc. Technology of Rolling Production (in Russian). Vol. 1. Metallurgy, Moscow, 1991.
4. Henry Jewik, R. P. Stratford, Thomas, C. W. Torque Amplification and Torsional Vibration in Large Mill Drives. *IEEE Transactions on Industry and General Applications*, Vol. IGA-5, No. 3, May/June 1969.
5. Perelmuter, V. M. *Electrotechnical Systems, Simulation with Simulink and SimPowerSystems*. CRC Press, Boca Raton, FL, 2012.
6. Vydrin, V. N. Dynamics of Rolling Mills (in Russian). Metallurgizdat, Sverdlovsk, 1960.
7. Chattopadhyay, A. K. Alternating Current Drives in the Steel Industry. *IEEE Industrial Electronics Magazine*. Vol. 4, No. 4, December 2010.
8. Perelmuter, V. Advanced Simulation of Alternative Energy, Simulation with Simulink and *SimPowerSystems*. CRC Press, Boca Raton, FL, 2020.
9. Voronin, S. S., Gasiyarov, V. R. The Development of the Electromechanical Screw-Down Mechanism Motion Control System of the Hot Plate Rolling Mill. 2016 IEEE NW Russia Young Researchers in Electrical and Electronic Engineering Conference (EIConRusNW).
10. Voronin, S. S., Gasiyarov, V. R. Modernization of the Hot Plate Rolling Mill Thickness Control System Using Hydraulic Gap Control Cylinders. 2017 IEEE II International Conference on Control in Technical Systems (CTS), St. Petersburg, Russia.
11. Radionov A. A., Maklakov A. S., Gasiyarov V. R. Smart Grid for Main Electric Drive of Plate Mill Rolling Stand. 2014 International Conference on Mechanical Engineering, Automation and Control Systems (MEACS). Tomsk, Russia.
12. Khramshin, V. R., Gasiyarov, V. R., Karandaev A. S., Baskov, S. N., Loginov, B. M. Constraining the Dynamic Torque of a Rolling Mill Stand Drive, *Bulletin of the South Ural State University*. Ser. Power Engineering. Vol. 18, No. 1, pp. 101–110, 2018. DOI: 10.14529/power180113.
13. Pittner, J, Simaan, M. A. An Initial Model for Control of a Tandem Hot Metal Strip Rolling Process. *IEEE Transactions on Industry Applications*, Vol. 46, No. 1, January/February 2010.
14. Voskanyants, A. A. Automatic Control of Rolling Processes (in Russian). Bauman MSTU, Moscow, 2010.
15. Zhong, Z., Wang, J. Looper-Tension Almost Disturbance Decoupling Control for Hot Strip Finishing Mill Based on Feedback Linearization. *IEEE Transactions on Industrial Electronics*, Vol. 58, No. 8, August 2011.
16. Pittner, J., Simaan, M. A Control Method for Improvement in the Tandem Hot Metal Strip Rolling Process. 2010 IEEE Industry Applications Society Annual Meeting. Houston, TX.
17. Pittner, J., Simaan, M. An Optimal Control Method for Improvement in Tandem Cold Metal Rolling. 2007 IEEE Industry Applications Annual Meeting, New Orleans, LA.
18. Pittner, J., Simaan, M. Use of Advanced Control with Virtual Rolling to Improve the Control of the Threading of the Tandem Hot Metal Strip Mill. 2016 IEEE Industry Applications Society Annual Meeting, Portland, OR.
19. Perelmuter, V. M., Sidorenco, V. A. DC Thyristor Electrical Drives Control Systems (in Russian). Energoatomizdat, Moscow, 1988.
20. Tselikov A. I., Polukhin P. I., Grebenik V. M. et al. Machines and Units Metallurgical Plants. V. 3. *Machines and Units for the Production and Finishing of Rolled Products (in Russian)*. Metallurgy, Moscow, 1988.
21. Perelmuter, V. M. Three-Level Inverters with Direct Torque Control. IEEE-IAS-2000. *Conference Record*, Vol. 3, 1368–1374, Rome, Italy, 2000.
22. Xin, Z., Dian-hua, Z., Ping-wen, C., Xu, L., Jie, S. On Quality Control Strategy in Last Stand of Tandem Cold Rolling Mill. 31st Chinese Control Conference, July 2012, Hefei, China.

23. Druzhinin, N. N. *Continuous Mills as an Object of Automation (in Russian)*. Metallurgy, Moscow, 1975.
24. Konovalov Yu., V. Roller handbook. In Book 2. *Production of Cold-Rolled Sheets and Strips (in Russian)*. Heat Engineering, Moscow, 2008.
25. Perelmuter V. M., Chelombitko A. V. Experimental Studies of Continuous Cold Rolling Mills as Objects of Regulation (in Russian). *Metallurgical and Mining Industry*. No. 3, 1969, 22–25.
26. Bryant, G. F., Butterfield, M. H. Simulator Assessment of Tandem Cold-Rolling-Mill Automatic Gauge-Control Systems. *Proceedings IEE*, Vol. 111, No. 2, February 1964.
27. Tezuka, T., Yamashita, T., Sato, T., Abiko, Y., Kanai, T., Sawada, M. Application of a New Automatic Gauge Control System for the Tandem Cold Mill. *IEEE Transactions on Industry Applications*, Vol. 38, No. 2, March/April 2002.
28. Pittner, J., Simaan, M. A., Samaras, N. Improved Threading of the Tandem Cold Mill sing Advanced Control Techniques with Virtual Rolling. 2019 IEEE Industry Applications Society Annual Meeting. Baltimore, MD, USA.

2 Simulation of the Paper-Making Machines

2.1 GENERAL ISSUES

A paper-making machine (PM) is a complex technological unit containing a large number of electric drives of various powers interconnected by a web of paper. The PM consists of several sections that perform various technological functions, while the web is simultaneously in all sections, so that the speed of its passage through each section should be approximately the same. A little speed difference arises only due to a change in the physical properties of the web as it passes through the next section. From a constructive point of view, each electric drive is a roll or several rolls interacting with the web either directly or through intermediate means (fabric, felt). Usually, PM consists of the following sections: forming, press, dryer, calender, and reel.

There are many types and varieties of PM, but from the point of view of the electric drive, their requirements are more or less the same, which allows us to consider a relatively small number of models of electric drives. In addition to the PM itself, the paper production plant may also include other installations with adjustable electric drives, such as supercalender, paper cutting machines, coating machines, etc.

To select the power of motors of PM electric drives, the specific power method is most often used: for each PM mechanism, the specific power (normal load, p_n) is determined expressed in kW per 1 m of the web width at a speed of 100 m/min. The corresponding tables are given in [1] (see also [2]). Since there specific powers are given in hp per 1 inch of width at a speed of 100 f/min, then to convert to the CI system, the given values must be multiplied by $0.746 \times 1000/25.4/0.3048 = 96.4$. For example, $p_n = 0.03$ for the transfer press, then, for the PM, with the web width of 4.8 m and a speed of 762 m/min, the expected power of the section motor is $P = 0.03 \times 96.4 \times 4.8 \times 7.62 = 106$ kW.

When choosing electric motors for new PMs, it is recommended, instead of p_n values, to use RDC values (Recommended Drive Capacity), which are 20%–30%, and sometimes more, higher than p_n values. It is assumed that the electric drive must cope with such an increased load in any case. This value takes into account gear losses, as well as, in some cases, the necessary dynamic torque. It must be taken into account that the selected motors must provide the starting torque of the section, which is usually 2–2.5 of the rated torque. Some details of the calculation of the section drive motor torques are given below when they are simulated.

The general features of PM electric drives are as follows. The paper has little tear resistance and is easy to fold. Therefore, the established ratios of motor rotational speeds must be kept with high accuracy; it is considered that it is not worse than 0.1%, which entails the mandatory use of digital sensors and rotational speed controllers. Moreover, in contrast to the rolling mills considered in Chapter 1, the moment of inertia of the load in PM can greatly exceed the moment of inertia of the motor, which causes

DOI: 10.1201/9781003394419-2

elastic vibrations in the system and makes it difficult to ensure the high accuracy of the control system, especially in the presence of backlash in the gears, and requires the use special measures, to which extensive literature is devoted [3–6]. It should be noted that a new trend in PM electric drives is the use of permanent magnet synchronous motors (PMSM), which are connected to the driven shafts without a gearbox, which eliminates backlash and increases the rigidity of the mechanical system [7].

The block diagram of the electric drive model taken during investigations is shown in Figure 2.1. A two-mass mechanical system is considered, which is described by equations (1.12)–(1.14). The index m refers to the motor and L to the working shaft (to the load). All variables are reduced to the axis of the working shaft, as is customary in [3–5]. The delay in the response to the reference motor torque is modeled by the first order element with the time constant T_μ, as it has been repeatedly used before. It is assumed that the measurement of the motor rotational speed is carried out digitally using a pulse encoder for a time interval T_i. The measurement is carried out with some error, which can be taken as a random variable uniformly distributed in the interval $\pm a$, where a is the sensor resolution.

If to define as

$$\omega_r = \sqrt{C\frac{J_m+J_L}{J_m J_L}}, \omega_{ar} = \sqrt{\frac{C}{J_L}} \tag{2.1}$$

resonant and antiresonant frequencies, respectively, the relationship (2.2) can be written for the motor rotational speed (for $T_L = 0$)

$$\omega_m = \frac{1}{(J_m+J_L)s}\frac{T_m(1+2\zeta_z s/\omega_{ar}+s^2/\omega_{ar}^2)}{1+2\zeta_n s/\omega_r+s^2/\omega_r^2} \tag{2.2}$$

The ratio of the rotational speeds of the load and motor is

$$\frac{\omega_L}{\omega_m} = \frac{1+2\zeta_z s/\omega_{ar}}{1+2\zeta_z s/\omega_{ar}+s^2/\omega_{ar}^2} \tag{2.3}$$

Here

$$\zeta_n = \frac{B\omega_r}{2C}, \zeta_z = \frac{B\omega_{ar}}{2C} \tag{2.4}$$

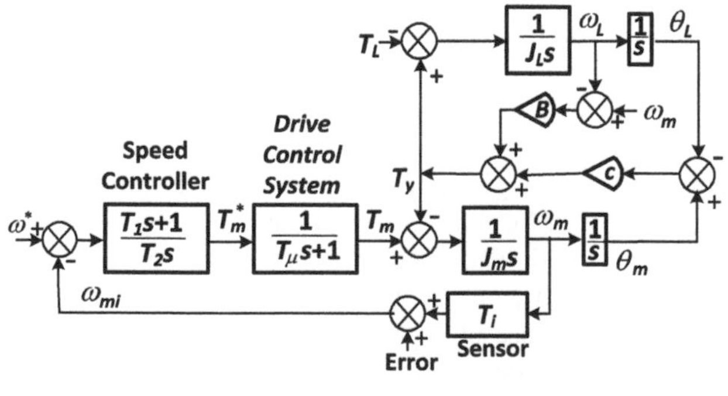

FIGURE 2.1   Block diagram of the electric drive model taken during PM investigations.

Some data related to the estimates of the elasticity of various sections of the PM are given in [3]. Since complete data are not given there, in the following models these data are used for the initial estimation. For **Paper1** model, as for the press section, it is taken $J_L/J_m = 14,7, J_m = 67$ kg-m^2, C = 10^5, т.е. $\omega_r = 40$s^{-1}.

The value of the coefficient B in (1.12) is difficult to calculate and can only be found empirically. So, for example, in [6], as a result of studying modern (at that time) PM in the USSR, it was found that the damping coefficient ζ_n for different sections varies in the range of 0.04–0.11 when ω_r changes in the range of 40–80 rad/s. Taking the average value $\zeta_n = 0.07$, one receives $B = 350$.

The above is implemented in the **Paper1** model. Since the speed digital measurement over time T_i actually measures the average speed over this time, the meter can be modeled as a **Mean** block that continuously measures the average value of the input variable in a running window of width T_i, at the output of which a storage block is set with a sample time T_i. It is taken that a steady-state error of the speed measurement should be 0.01%. The more the allowed measurement time, the easier it is to achieve this accuracy [8]. Therefore, when modeling the system, it is reasonable to determine the dependence of the quality of processes in the system on the value of T_i. In the **Paper1** model, the T_i value is given in the field *Model Properties/Callbacks/InitFcn*. At this, the transfer function of the speed controller changes, which, for each value of T_i, was chosen so as to obtain an overshoot of 10%–15% with a rise time of about 1 s.

This model provides for the possibility of reducing the oscillation of the system by introducing an acceleration feedback signal. Figure 2.2 shows the changes in the motor rotational speed and motor torque with $T_i = 5$, 10, and 15 ms. It can be seen that with increasing T_i, the minimum achievable overshoot increases. It should be

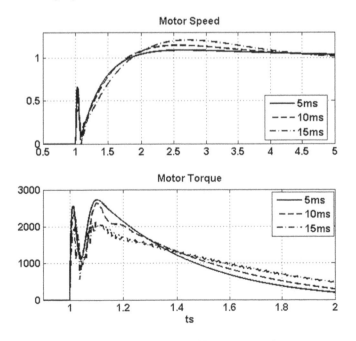

FIGURE 2.2 Model **Paper1** step response for different values of T_i.

noted that the process at $T_i = 15$ ms was obtained with the introduction of an acceleration feedback, since otherwise the process is highly oscillatory, while the other two processes were obtained without this feedback.

Figure 2.3 shows the change in the shaft rotational speed and the elastic torque under the load surge. It can be seen that, due to the large moment of inertia of the working shaft, the processes proceed very smoothly, and, with the selected parameters of the controllers, the maximum speed drop decreases with Ti decrease.

The model provides the opportunity to simulate the error in measuring the rotational speed caused by an amplitude quantization. In the **Uniform Random Number** block, limiting values are set equal to the "weight" of one pulse (i.e., numbers $\pm\varepsilon$, where ε is the accepted measurement error, for example, $\varepsilon = 1/10,000$), and the parameter *Sample Time = T_i*. If to perform simulation, one can see that for the selected parameters, the deviations of the motor rotational speed will not exceed the value ε, oscillations appear in the motor torque, the numerical values of which can be estimated knowing the nominal parameters of a particular motor (power, rotational speed).

In the **Paper1b** model, the system parameters are close to those for the calender [4]: $J_L/J_m = 67.6$, $J_m = 28$ kg-m², $C = 300,000$ N-m/rad, that is, $\omega_r = 103.5\,\text{s}^{-1}$, $B = 406$ N-m-s/rad. Investigation of this model shows that only with $T_i = 5$ ms, with the acceleration feedback loop enabled, satisfactory processes can be obtained. When values of T_i are larger, the processes have either a large overshoot or (and) a large oscillation. Improving the quality of transients is achieved by introducing a series notch filter with a transfer function

$$W_f = \frac{\omega_f^2 + 2\zeta_f\omega_f s + s^2}{\omega_f^2 + 2\zeta_{fd}\omega_f s + s^2} \tag{2.5}$$

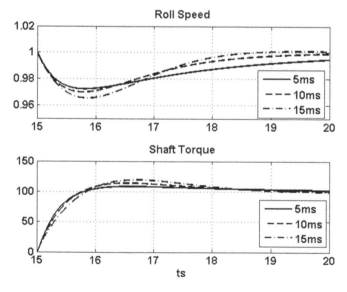

FIGURE 2.3 Model **Paper1** responses when the load surge.

where the parameters ω_f and ζ_f are as close as possible to the parameters ω_r and ζ_n of the system under study, and ζ_{fd} corresponds to a well-damped system, for example, $\zeta_{fd} = 3$, model **Paper1b1**. The resulting processes are shown in Figure 2.4. It can be seen that they are quite satisfactory, the overshoot does not exceed 10%, and the settling time is 1 s, but in reality, the parameters ω_r and ζ_n of the system are not known exactly and can change during operation and over time. Figure 2.5 shows the process when the J_L, C, and B values change by 20% with the filter transfer function unchanged ($J_L/J_m = 78.3$, $J_m = 29$ kg-m^2, $C = 360{,}000$ N-m/rad, that is, $\omega_r = 112.2\,\mathrm{s}^{-1}$, $B = 327$). It can be seen that the processes in the system do not change significantly. Note that without this series correction, the system is unstable.

In the **Paper1c** model, the system parameters are close to those for the dryer section [4]: $J_L/J_m = 153$, $J_m = 493$ kg-m^2, $C = 600 \times 10^3$ N-m/rad, that is, $\omega_r = 35\,\mathrm{s}^{-1}$, $B = 2400$. To eliminate oscillations, a notch filter is used; without it, the system is unstable. The process is shown in Figure 2.6. It can be seen that the overshoot increased to 15%, and the settling time increased by 1.5–1.7 times. The reaction of the system to a change in its parameters with the same controllers is approximately the same as in the previous model.

Since the paper has a low strength, its tension when passing through the PM must be small and maintained with sufficient accuracy. It is supposed that the tension should be approximately 10% of the tensile strength [9]. It should be noted that in the production of paper, the tensile strength is not expressed in N/mm^2, as in rolling production, but in N/m or N/mm of the paper sheet width (the so-called Ultimate Tensile Strength), which, according to [10], is equal to $(2.8–3.7) \times 10^3$ N/m, so that with a paper web width of 4.8 m, the tension should be 1.3–1.8 kN.

The specified tension value is kept by appropriate control, with high accuracy, of the rotational speeds of the motors of adjacent driven rolls. However, in some

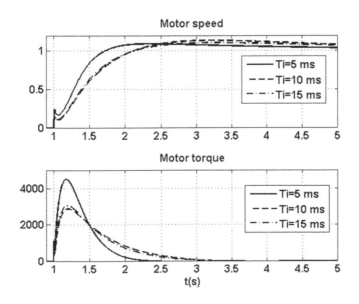

FIGURE 2.4 Processes in the PM drive with the notch filter.

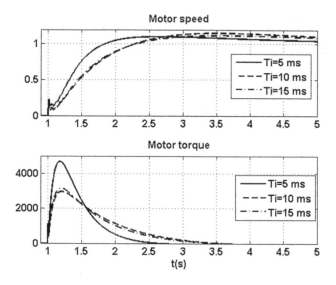

FIGURE 2.5 Processes in the PM drive with the notch filter when parameters of the PM can change.

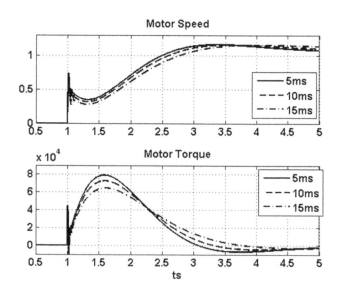

FIGURE 2.6 Processes in the PM drive with the large moment of inertia of the load.

cases, especially for the output sections of the PM (the so-called dry part), the accuracy achieved in this case is insufficient due to changes in both the elasticity of the processed material with changes in temperature and humidity, and circumferential speed of the rolls. The latter thing is due to the fact that most of the rolls used in PM have a soft coating, which is subject to the influence of external conditions, so that even at a constant rotational speed, the circumferential speed can change [9].

Therefore, at the most critical points of the PM, direct tension control is used using tension sensors or observers.

In relation to Figure 2.7, the change of the tensile strain ε_2 between adjacent rolls is described by the equation [10]

$$\frac{L}{1+\varepsilon_2}\frac{d\varepsilon_2}{dt} = v_2 - \frac{v_1(1+\varepsilon_2)}{1+\varepsilon_1} \tag{2.6}$$

where ε_1 is the strain in the previous span, v_1, v_2 are the roll tangential velocities. In turn, the tension $T_2 = QE\varepsilon_2$ where Q is the web cross-section area, E is the modulus of elasticity which is $(0.5–1) \times 10^7$ kN/m² depending on the grade and humidity of the paper. In a number of researches, the dependence of ε_2 on ε_1 is neglected. Then, assuming that the speeds v_1 and v_2 do not differ significantly from the average speed of the PM V_0, the equation is obtained that resembles (1.31) by form:

$$\tau\frac{dT_2}{dt} + T_2 = QE\frac{v_2-v_1}{V_0} = K_s\frac{v_2-v_1}{V_0} \tag{2.7}$$

where $\tau = L/V_0$. It is taken hereinafter $Q = 4.8\,\text{m} \times 51\,\mu\text{m} = 245 \times 10^{-6}\,\text{m}^2$, $V_0 = 12.7\,\text{m/s}$, $L = 4\,\text{m}$, so that $\tau = 0.31$ s, the factor K_s varies within 1225–2450 kN, and to receive the tension of 1.5 kN, it must be $\Delta V/V_0 = (0.05–0.1) \times 10^{-2}$.

These relations are applied to the joint operation of the end dryer section and the calender, the parameters are given in [10]. The diameter of the drum of the dryer section is 1524 mm, so the rotation speed of its axis is $12.7 \times 2/1.524 = 16.7$ rad/s. The rated torque of the motor, reduced to the axis of the drum, with a gearbox with $i = 12.4$ is $1170 \times 12.4 = 14.51$ kN. The diameter of the calender roll is 600 mm, so the speed of rotation of its axis is 42.3 rad/s. The rated torque of the motor, reduced to the shaft axis, with a gearbox with $i = 5.06$ is $520 \times 5.06 = 2.63$ kN.

This system is simulated in the **Paper1d** model. The models of the dryer section and calender are taken from the **Paper1c** and **Paper1b1** models respectively. The relation (2.6) and the tension T_2 are computed in the **Tension** subsystem. The quantity ε_1 is taken as an independent variable, knowing it, the tension T_1 is computed. T_1 and T_2 values are used to calculate the motor load torques as

$$M_D = M_{D0} + R_D(T_1 - T_2) \tag{2.8}$$

$$M_c = M_{c0} + R_c T_2 \tag{2.9}$$

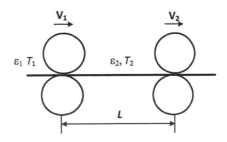

FIGURE 2.7 For calculation of the tension of the paper web.

Here R_D and R_C are the radiuses of the rolls of the dryer section and calender respectively, whose nominal values are 0.762 and 0.3 m, M_{D0} and M_{C0} are the motor load torques, whose estimated values, taking into account the data given in [9,10], are taken equal to 34%–40% of the motor rated torques. Remind that all values are reduced to the working shafts, not to the motor axes.

The rate of change of the PM given speed V_0 is limited by the block **Rate Limiter**; the acceleration rate is set much higher than the actual values in order to reach quickly the steady state, in which the operation of the tension controller is simulated. After dividing by the nominal values of the radiuses of the working roll or drum, the references are obtained for speed controllers of electric motors in rad/s. After multiplying the real rotational speed by the actual radius, which, for the reasons mentioned above, may differ from the nominal one, a signal of the circumferential speed of the drum (working roll) is generated. The calender rotational speed reference is increased by $\Delta\omega$ to create the required tension before turning on the controller.

Tension controller is PI type. It is put into operation at $t = 120$ s, when it is assumed that the steady state or close to it has been reached already. The controller affects the rotational speed of the calender motor. Drive models have the opportunity to simulate fluctuations caused by speed digital measurement errors. This factor is especially important when the tension change is investigated, since even a small deviation of the speed from the given one causes a significant change in tension. For example, the speed error of 0.01% causes the tension change $\Delta T = 245 \times 10^{-6} \times 10^{10} \times 10^{-4} = 245$ N, that is, 16% of the given value.

Some simulation results are given below. Figure 2.8a shows the process of turning on the tension controller, after which the wanted tension value is set to 1500 N. and subsequent changes to 1800 N at $t = 150$ s and back to 1500 N at $t = 180$ s. It can be seen that references are processed quickly and without significant overshoot. Figure 2.8b shows the process when the modulus of elasticity decreases by 20% at $t = 200$ s that leads to a decrease in tension, which is restored quickly. At $t = 230$ s, tension T_1 appears, corresponding to the value $\varepsilon_1 = 0.5 \times 10^{-3}$ that initially causes an increase in tension T_2, which then quickly recovers.

Figure 2.8c shows the process when, at $t = 250$ s, the drum radius of the dryer section decreases by 1% for 4 s, which leads to a noticeable increase in tension, which is then restored by the action of the controller. At $t = 270$, the load torque of the dryer section increases from 5 to 6 kN-m.

It should be noted that considered perturbations changed instantly or very quickly under the simulation, while in practice they change more smoothly, so that smaller values of the maximum deviations should be expected. Of course, such a high-quality regulation is due to the ideality of the tension meter. The problems associated with this are covered in [9].

The precision measurement of the two motor rotational speeds is shown in Figure 2.8. The plots in Figure 2.9 repeat the plots in Figure 2.8a, but they were obtained by taking into account the errors caused by the discreteness of the speed measurement by digital methods. In the upper plot, the error is 0.1%, and tension fluctuations are visible. In the lower plot, the error does not exceed 0.01%, tension fluctuations are not noticeable.

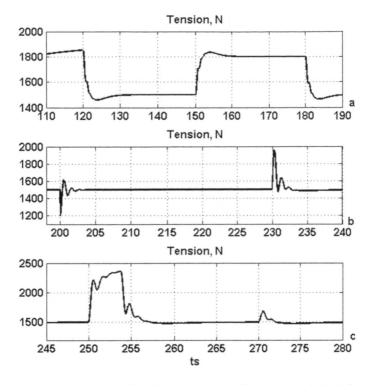

FIGURE 2.8 Tension changes (a) reference change; (b) modulus of elasticity and back tension change; (c) drum diameter change.

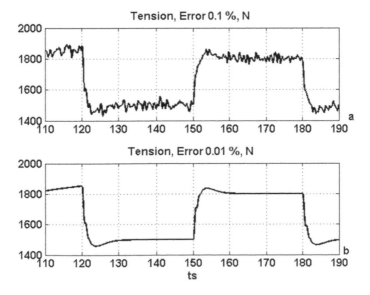

FIGURE 2.9 Tension change when the discreteness of measurement is taken into account.

2.2 MODELING SECTIONAL ELECTRIC DRIVES

2.2.1 Forming Section

The forming section is a main part of PM in which the paper web with water content from 18% up to 22% is created by dewatering. There are different designs of the forming section. The most widespread construction, with the flat forming table, is considered here. Its main component is a moving endless wire that is pulled taut between the breast and couch rolls. The top (working) wire part moves over a number of devices that carry out the removal of some of fluid, and the bottom (idle) part—over the turning roll and over several wire-leading, wire guide, and wire stretch devices. Among the rolls mentioned above, the couch roll plays the main function in web dehydration, while the function of the others is to move the wire. The wire turning and couch rolls, and in some PM, one wire guide roll, are driven, while the remaining rotating rolls, including the large-diameter breast roll, do not have their own motor and are driven by the wire. Figure 2.10 shows the arrangement of the main rolls.

When studying electric drives of the forming section, a number of features should be taken into account. Since many rolls are driven by the wire, it must be taut. It is supposed that the wire made of synthetic material must be stretched with a force $q = 5–7$ kN per 1 m of width; for metal wires, lower values are acceptable. This tension is created by a special tension device placed at the bottom part of the wire. The torque created by the drive shaft must not exceed the values determined by the friction forces between the roll and the wire. This torque depends on the wrap angle of the roll by the wire α and the coefficient of friction between them μ. The power corresponding to this torque can be determined by the formula

$$P_{wtm} = q\left(e^{\mu\alpha} - 1\right)Wv \tag{2.10}$$

where W is the wire width, m, v is PM speed m/s. The tensions before q_{in} and after q_{out} the roll are coupled by the relation

$$q_{in} = q_{out}e^{\mu\alpha} \tag{2.11}$$

Where μ is the friction coefficient of the synthetic wire on the rubber coating, $\mu = 0.25 \div 0.45$ depending on rubber hardness; $\alpha = 2.5–2.8$ rad for wire turning roll. Taking, for example, $q = 6$ kN/m, $\mu = 0.3$, $\alpha = \alpha_{wt} = 2.8$ rad, $W = 4.8$ m, $v = 762$ m/min $= 12.7$ m/s, one finds $P_{wtm} = 6\times\left(e^{0.3\times2.8} - 1\right)\times 4.8\times 12.7 = 482$ kW.

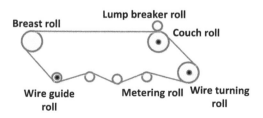

FIGURE 2.10 Arrangement of the main rolls of the forming section.

The wrap angle is small for the couch roll, so the torque value computed by (2.10) is also small, but, after the vacuum is set and the presser is put down, the maximal transferred torque of the couch roll sharply increases. Computation of this value which is designated as ΔP_c demands the detailed knowledge of the couch roll construction [11]. It can be taken approximately

$$\Delta Pc = p_{ck} b_c \mu v W, \qquad (2.12)$$

where p_{ck} is the average vacuum value in the suction chamber, Pa, b_c is the width of this chamber, m. Let, for example, $p_{ck} = 68\,\text{kPa}$, $b_c = 0.25\,\text{m}$, then $\Delta P_{ck} = 68 \times 0.25 \times 0.3 \times 12.7 \times 4.8 = 311$ кВт. Since q_{in} for turning roll is q_{out} for the couch roll, for the letter, by wrap angle $\alpha_c = 0.68\,\text{rad}$,

$$P_{cm} = qe^{\mu \alpha_{wl}}\left(e^{\mu \alpha c} - 1\right)Wv = 6 \times e^{0.3 \times 2.8} \times \left(e^{0.3 \times 0.68} - 1\right) \times 4.8 \times 12.7 = 108.8 \;\; \text{кВт, т.е.}$$

$\Delta P_c \gg P_{cm}$. The difference will be even greater if to take into account the pressing force of the presser.

The given fact leads to the use of an electric drive system with a variable structure: before the start of the normal operation of the PM, the electric drive of the couch roll operates in the speed controller mode with a soft characteristic and reduced current limitation, and after the vacuum is established, the electric drive operates in the torque controller mode, and the torque setting is determined as part of the torque setting of the turning roll drive. The values of the maximum transmitted torques can be used to select motors and to distribute the load between them. The power distribution between the couch roll and the wire turning roll varies from (30:70)% to (50:50)%. When using synthetic fabrics instead of metal fabrics, 30%–40% more power is required.

When considering the dynamic properties of the electric drives of the forming part, it should be borne in mind that the motors of the drive rolls (we will take into account only the turning and couch roll) must provide the required dynamic torque also for non-drive rolls rotating due to the connection with the wire, that is, the wire with all rolls associated with it should be considered as a single inertial system. The transfer of torque between the rolls is carried out by the change of the elongation of the wire in different areas. To take into account this phenomenon correctly, it is necessary to know the location of the rolls and the modulus of elasticity of the wire material, which is not always possible. Therefore, the simulation results presented below give, to a certain extent, approximate information.

Firstly, the motors for the driven rolls are chosen. For the sum of the turning and couch rolls, $p_n = 0.086$; taking $p_{rdc} = 1.3\,p_n$ and taking into account power increase by the factor 1.5, because of employment of the synthetic wire, the power of the motors is $P = 0.086 \times 96.4 \times 1.3 \times 7.62 \times 4.8 \cdot 1.5 = 591$ kW. The powers of the motors of the turning and couch rolls are taken as 325 and 280 kW respectively, with the voltage of 690 V and $Z_p = 3$. The motor parameters turning roll/couch roll are nominal rotational speed 992/991 rpm, efficiency = 96.1/95.9, cos φ = 0.8/0.79, $I_{nom} = 362/309$ A, the no-load current 170/138 A, the starting current ratio 5.3/5, the nominal torque 3130/2698 N-m, the starting torque ratio 0.8, the maximum torque ratio 2.1/2, $J_m = 5.7/5$ kg-m^2.

The diameter of the turning roll is taken equal to 0.8 m and the mass of 7630 kg. The shaft rotational speed is $\omega_{wt} = \dfrac{12.7}{0.4} = 31.75$ rad/s or 303 rpm. The shaft is driven with the gearbox $I = 3.2$. Neglecting the small thickness of the roll wall, its moment of inertia is $J_L = 7630 \times 0.16 = 1221$ kg-m^2. Thus, $J_L/J_m = 1221/(5.7 \times 3.2^2) = 20.9$. The stiffness coefficient C is calculated with use of (1.15), (1.16) at $T_{cr} = 2P_{wtm}/\omega_{wt} = 2 \times 325/31.75 = 20.5$ kN-m, $\tau = 80$ N/mm^2, $l = 3$ m:

$$D = \sqrt[3]{\frac{16 \times 20.5}{\pi \times 80 \times 10^3}} = 0.11 \text{ m}, C = \frac{\pi \times 80 \times 0.11^4 \times 10^9}{32 \times 3} = 0.38 \times 10^6 \text{ N-m/rad},$$

that is, $\omega_r = \sqrt{\dfrac{0.38 \times 10^6 \times (1221 + 58.4)}{1221 \times 58.4}} = 82.6$ rad/s.

The diameter of the couch roll is taken equal to 1 m with the mass of 12,700 kg. The rotational speed of its shaft is $\omega_c = \dfrac{12.7}{0.5} = 25.4$ rad/s or 243 rpm. The shaft is driven with the gearbox $i = 4$. Neglecting the small thickness of the roll wall, its moment of inertia is $J_L = 12,700 \times 0.25 = 3175$ kg-m^2. Thus, $J_L/J_m = 3175/(5 \times 4^2) = 39.7$. The stiffness coefficient C is calculated with use of (1.15), (1.16) at $T_{cr} = 2(P_{cm} + \Delta P_c)/\omega_c = 2 \times 420/25.4 = 33.1$ kN-m, $\tau = 80$ N/mm^2, $l = 3$ m:

$$D = \sqrt[3]{\frac{16 \times 33.1}{\pi \times 80 \times 10^3}} = 0.13 \text{ m}, C = \frac{\pi \times 80 \times 0.13^4 \times 10^9}{32 \times 3} = 0.75 \times 10^6 \text{ N-m/rad},$$

that is, $\omega_r = \sqrt{\dfrac{0.75 \times 10^6 \times (3175 + 80)}{3175 \times 80}} = 98$ rad/s.

The diameter of the non-driven breast roll is 0.8 m and the mass of 3244 kg, that is, under the same assumptions, its moment of inertia is $J_b = 3244 \times 0.16 = 519$ kg-m^2.

The **Paper2** model uses the DTC system **AC4** from **Simscape** library, to drive the rolls. In this model, the DC link of the inverter is formed by a diode rectifier, so recuperation is impossible, but this mode is not typical for the PM. When a DC overvoltage occurs, a braking resistor is activated to limit the overvoltage. The resistances and inductances of the IM are chosen in such a way that the characteristics of the motor model are sufficiently consistent with the above characteristics of the simulated IM. For this purpose, program `ID _ PARM1` is executed.

Program `ID _ PARM1`

```
V=690;
Inom=362;
Kss=5.3;
Is=Inom*Kss;
Ll=V/1.732/Is;
Lsl=Ll/2/314
Lrl=Lsl
Ix=170;
Lx=V/1.732/Ix/314;
```

```
Lm=Lx-Lsl
ks=Lm/Lx;
mm=2.1;
sn=0.008; x=mm+sqrt(mm*mm-1)
sk=sn*x
xr=(Lsl+ks*Lrl)*314
Rr=sk*xr
```

This program does not determine the value of the stator resistance, which is assumed to be approximately equal to the average value (in p.u.) for various IMs of approximately the same power.

When the **AC4** model was applied, some changes were made in it (after executing the *Break Link* command): a block for measuring the rotational speed from previous PM models with a measurement time of 10 ms was added and the possibility of using a notch filter was provided. Elements simulating a two-mass system, borrowed from previous PM models, have been added to the model diagram, but they additionally take into account the presence of a gearbox between the motor and the roll.

To reduce the simulation time, the acceleration rate is set much higher than the real one, but if it is necessary to study this mode, it can be easily changed. When modeling, it is assumed that at $t = 15$ s, a steady state is achieved at the rotational speed of 900 rpm and with a motor torque of 3 kN-m.

Figure 2.11 shows the processes with a step change of the speed reference from 900 to 950 rpm and with a step change in the load torque by 5%, from 3 to 3.15 kN-m.

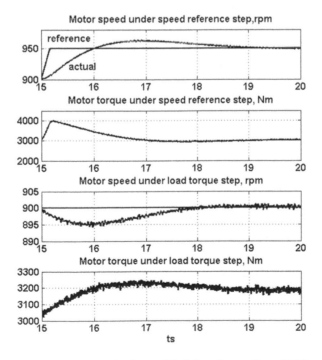

FIGURE 2.11 Dynamic processes in the model of the wire turning roll drive.

The notch filter has been disabled. It can be seen that the rise time in the first case and the maximum speed deviation time in the second case are 1 s, and the duration of the processes (setting time) is 3 s.

Together with DTC, vector control systems are used. As it is known [12], there are systems of direct and indirect vector control. The latter are simpler, but require accurate knowledge of the IM parameters, which are not always known and can change during operation. Since high requirements for the quality of regulation are made of demand for PM electric drives, direct vector control systems are used for PM drives. When using such a system, it is necessary to determine the modulus and position of the rotor flux vector. For this purpose, firstly, by measuring the voltage and current of the stator, using relations (1.17)–(1.19) or similar, the components of the stator flux linkage vector $\Psi_{s\alpha}$, $\Psi_{s\beta}$ are calculated, then, by the relations

$$\Psi_{r\alpha(\beta)} = \frac{L_r}{L_m}\Psi_{s\alpha(\beta)} - \sigma L_s i_{s\alpha(\beta)} \qquad (2.13)$$

$$\sigma = 1 - \frac{L_m^2}{L_s L_r}, \qquad (2.14)$$

the components of the rotor flux vector Ψ_r are computed, whose modulus and angle are calculated by the relationships that are analogous to (1.20).

In the control system, the components I_d, I_q of the space vector of the stator current are controlled; the first component, directed along to the vector Ψ_r, determines the value of the Ψ_r modulus and therefore its set value is determined by the output of the controller of the Ψ_{rmod}; the second component is directed perpendicular to the vector Ψ_r and determines the magnitude of the motor torque, and therefore its set value is determined by the output of the speed controller. Since the control systems of the currents I_d, I_q turn out to be coupled, to decouple them, the signals $-\omega_m\sigma L_s I_q^*$ and $\omega_m L_s I_d^*$ are added to the outputs of the current controllers I_d, I_q respectively.

The block diagram of the control system is shown in Figure 2.12. Such a system is modeled in the **Paper2a** model. It is based on the previous model with modifications. The **AC4** subsystem is replaced with the new **IM_Drive** subsystem. Motor parameters are set in the Modeling/Model Setting/*Model Properties/Callbacks/InitFcn* field, since their values are also used in the control system. In the **IM_Drive** subsystem, the **DTC**

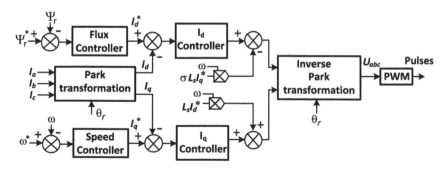

FIGURE 2.12   Block diagram of the direct vector control system.

subsystem is replaced by the **Observer** subsystem, in which the modulus and angle of the rotor flux vector are calculated using formulas (1.17)–(1.20), (2.13), and by the **Curr_Reg** subsystem, in which the stator current is regulated in accordance with the structure in Figure 2.12. If to run simulation in this model, one can see that the processes are practically the same as those in the previous model, so they are not shown here.

The **Paper2b** model is the same as the **Paper2** model, but the motor and load parameters correspond to the given parameters of the couch roll. This drive has a less favorable ratio between the moment of inertia of the load and the motor power than the one discussed above, so it was necessary to reduce the rate of its acceleration (in the PM, the rate of acceleration is determined by the possibility of the electric drive, which has the smallest maximum allowable rate of acceleration), and also change the parameters of the speed controller. Processes under the reference speed step change from 900 to 950 rpm and under the load torque step change by 5%, from 2.5 to 2.625 kN-m, are shown in Figure 2.13. The notch filter was deactivated as before.

The vector control method is used in the **Paper2c** model. Processes are very close to those shown in Figure 2.13.

In the following models, the joint operation of electric drives of the wire turning and couch rolls is investigated. In the **Paper2d** model, electric drives are modeled in a simplified way, as in the **Paper1-Paper1c** models. When creating a model, it is necessary to determine the principle of control of the load distribution. It is assumed in the model under consideration that after reaching the

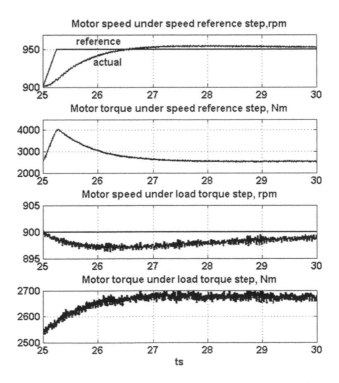

FIGURE 2.13 Dynamic processes in the model of the couch roll drive with DTC.

operating speed, before applying the load, the speed controller in the couch roll control system is turned off, and the torque setting is proportional to the torque setting (output of the speed controller) in the turning roll control system, with the proportionality factor K_{sh}^*. At the same time, the torque setting of the turning roll is reduced in proportion to the factor $1 - K_{sh}^*$. The total load of both electric drives is equal to the sum of the static load P_Σ, which is spent to overcome the resistances to move the wire and losses in gears, and the dynamic load that occurs when the speed of the wire changes and depends on the total moment of inertia of the non-drive rolls and the mass of the wire with the substance located on it and acceleration values. This load is distributed between the drives in proportion to the actual values of the torques of the electric drives. During operation, the circumferential speeds of the rolls can differ only by a very small amount, since they are coupled by a rigid wire. In the model, to take into account this fact, an action on the couch roll load torque is carried out; the action signal is in proportion to the difference of the circumferential speeds of the turning and couch rolls, with the large proportionality factor K. The structure of the model is shown in Figure 2.14.

It is taken under simulation, in accordance with p_n value, $P_\Sigma = 300\,kW$ that corresponds to the peripheral force $F_{st} = 300/12.7 = 23.6$ kN. The dynamic peripheral force is $F_{dyn} = mdV/dt$ where, with the breast roll mass of 3244 kg, the total mass of the non-driven wire parts is taken equal to 5000 kg. To reduce the simulation time, the acceleration rate is assumed to be much higher than the real one: $1\,m/s^2$. It is assumed that the wire is accelerated firstly to 20% of the rated speed, and then to full speed with ½ of the full load, while the turning roll provides 70% of the total load. At $t = 25$ s, the setting of the load distribution is activated, at which the couch roll creates 40% of the power, and at $t = 30$ s, a vacuum is created and the load increases to full. At $t = 40$ s, the wire speed increases by 5%. It can be seen (Figure 2.15) that after setting the ratio of torques, the specified ratio is kept: the torque of the wire

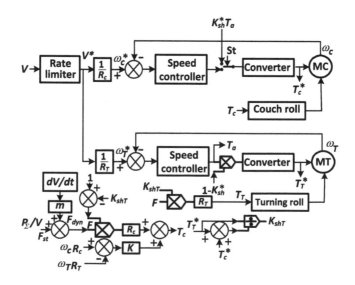

FIGURE 2.14   The structure of the **Paper2d** model.

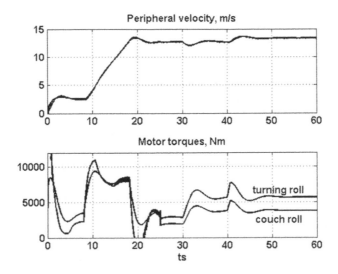

FIGURE 2.15 Processes in the **Paper2d** model.

turning roll is 5650 N-m, and of the couch roll is 3780 N-m, that is, the load of the latter is $3780/(3780+5650) = 0.4$. It should be noted that here all the torques are reduced to the working shafts.

Another variant of load distribution control is used in the **Paper2d1** model which is taken partially from [5]. The distribution of the load is carried out by a corresponding effect on the rotational speed of the couch roll motor: when it increases, the tension before the couch roll increases, which leads to an increase in its load, and the tension after the couch roll decreases, which leads to a decrease of the load of the wire turning roll. Change of the loads is carried out due to very small differences in the peripheral speeds of these rolls. The block diagram of the model is shown in Figure 2.16.

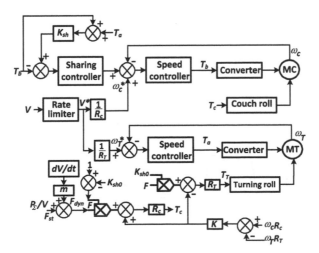

FIGURE 2.16 The structure of the **Paper2d1** model.

Figure 2.17 shows the process when, at $t = 25$ s, the load distribution control is turned on, when the couch roll creates 40% of the power, and at $t = 35$ s, a vacuum is created and the load increases to full. At $t = 50$ s, the share of couch roll increases to 45%. It can be seen that the specified load ratio is kept: at $t < 50$ s, the torque of the turning roll is 6150 N-m, and the couch roll is 4100 N-m, that is, the load of the latter is 4100/10,250 = 0.4, and at $t > 50$ s the load of the couch roll is 4710/(4710 + 5700) = 0.45.

The **Paper2e** model is the **Paper2d1** model, in which, instead of simplified models, the full electric drive models for turning and couch rolls are used; these models were used in the **Paper2** and **Paper2b** models respectively. Both models are powered from a common AC voltage source of 800 V, which in practice is a transformer that supplies all PM electric drives with this voltage level. In the model under consideration, at $t = 25$ s, after speeding up at first to the speed of 20% of the operating and afterward to the full one, the circuits that control the load distribution are activated, with the assigned value $K_{sh} = 0.4$; at $t = 30$ s, the vacuum and the full load of the forming section are created. Process is shown in Figure 2.18. It can be seen that after the operating condition is established, the load distribution corresponds to the required one: the torque of the turning roll (here the torques are reduced to the motor shafts) is 1800 N-m, and the couch roll is 1200 N-m, that is, the share of the couch roll is indeed 1200/(1200 + 1800) = 0.4.

2.2.2 PRESS SECTION

In the press section, the structure of the wet paper web is compacted, its surface is given an appropriate smoothness, and moisture is removed by compression. The wet paper web is picked up from the wire section and transported by means of a felt through the press section, which consists of several press nips formed by two rolls

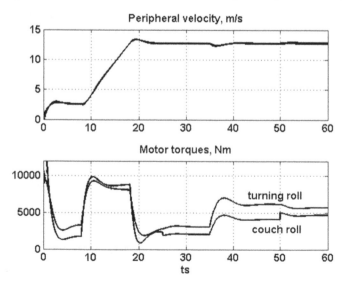

FIGURE 2.17   Processes in the **Paper2d1** model.

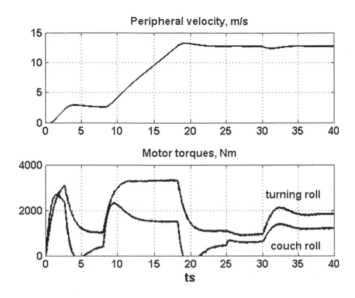

FIGURE 2.18 Processes in the **Paper2e** model.

pressed against each other. There are two and multi-roll presses. Figure 2.19 shows some options for the layout of the presses.

The press is modeled according to Figure 2.19b in the **Paper3** model. The lower roll is taken the same as the couch roll from the previous model: the roll diameter is 1 m, the moment of inertia is 3175 kg-m²; at the PM speed of 762 m/min, the roll rotational speed is 762/60/0.5 × (30/π) = 243 rpm The moment of inertia of a granite roll with a diameter of 0.8 m is taken equal to 2350 kg-m². As provided in [1,2], the normal load of such a press is equal to $p_n = 0.024 \times 96.4 \times 4.8 \times 7.62 = 86.4$ kW;

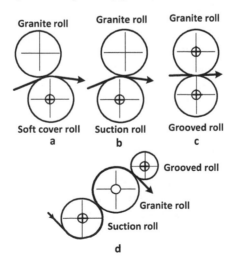

FIGURE 2.19 Some options for the layout of the presses (a) simple two-roll press; (b) two-roll press with the suction roll; (c) two-roll press with grooved roll; (d) three-roll press.

taking into account that for this press p_{rdc} is 1.3 times larger, as well as the need to provide a starting torque that can be 2.5 times higher than the normal load torque and a possible increase in the compression force of the rolls, the IM with a power of 160 kW at a voltage of 400 V, $Z_p = 3$ is taken, with the following data:

— rated speed	990 rpm
— cos φ	0.86
— efficiency	95.6
— rated current	280 A
— rated torque T_m	1543 N-m
— maximum torque ratio	2
— starting current ratio	6.8
— moment of inertia	10.22 kg-m²

The motor rotates the roll with the gearbox $i = 4$. The load total moment of inertia reduced to the bottom shaft is $J_L = 3175 + 2350 \times (1/0.8^2) = 7000$ kg-m² and the total moment of inertia is $J = 7000 + 10.22 \times 4^2 = 7164$ kg-m².

The stiffness coefficient C is computed by (1.15), (1.16) under $T_{cr} = 3T_m = 3 \times 1543 \times 4 = 18.5$ kN-m, $\tau = 80$ N/mm², $l = 3$ m:

$$D = \sqrt[3]{\frac{16 \times 18.5}{\pi \times 80 \times 10^3}} = 0.105 \text{ m}, C = \frac{\pi \times 80 \times 0.105^4 \times 10^9}{32 \times 3} = 0.32 \times 10^6 \text{ N-m/rad},$$

that is, $\omega_r = \sqrt{\dfrac{0.32 \times 10^6 \times 7164}{7000 \times 164}} = 44.7$ rad/s, $B = \dfrac{2 \times 0.32 \times 10^6 \times 0.07}{44.7} = 1002$ N-m-s/rad

Various conditions can be modeled in this model. The main one is the following: acceleration to a speed of 10% working one with a load of 2.5 working one, this speed is achieved approximately at $t = 3$ s; at $t = 3$ s, the roll pressing is set such that the total load becomes equal to 0.5 working load; at $t = 4$ s, acceleration to the operating speed of 972 rpm begins with the rate of 30 rpm/s. At the same time, the pressing force of the rolls begins to increase, so that the load increases from 0.5 working to working in 30 s (note that the working load reduced to the motor shaft is taken to be 900 N-m); at $t = 50$ s, the load increases by 22% in 2 s. To implement these conditions, several blocks have been added to the model. The **T_load** block controls the load. The **RL** subsystem is the rate limiter with the acceleration rate control by the extremal signal which is given by **Stair Generator ACC**. Acceleration at start-up is taken much less than the working one. Since the **AC4** model does not provide for the possibility of changing the acceleration during operation, the additional rate limiter is used for this purpose, which turns off after reaching the speed of 10%.

Figure 2.20 shows the processes in question. It can be seen how in the process of acceleration, due to the increase in the pressing force of the rolls, the load torque increases.

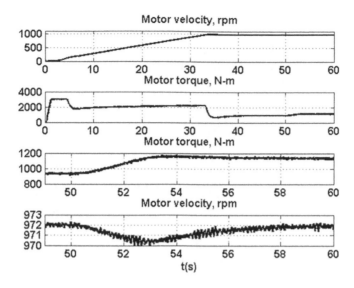

FIGURE 2.20 Processes in the **Paper3** model.

When the pressing force increases at the working speed, the motor rotational speed decreases by 0.2% and comes back to the initial value for 2 s.

To determine the quality of the system, such a "reference" mode is usually used, as a response to a step change of the speed reference. Figure 2.21 shows the process when the reference is stepped from 970 to 1000 rpm. Speed settling time is 1.2 s with 10% overshoot, the roll speed changes smoothly.

The press as in Figure 2.19c is modeled in the **Paper3a**, **Paper3a2** models. The upper granite roll is the leading one, its diameter is taken equal to 1 m with a moment

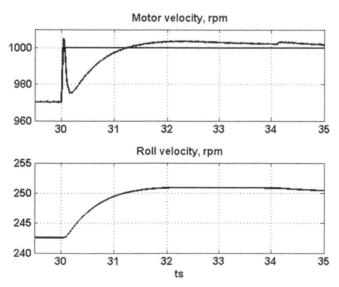

FIGURE 2.21 Press drive response to the reference step.

of inertia of 3680 kg-m², and the diameter of the lower roll is 0.8 m with a moment of inertia of 1800 kg-m². The PM speed is still equal to 762 m/min with a paper web width of 4.8 m. A press with such an arrangement can be used as the last press of a press group in the production of some grades of paper [5].

Unlike the previous model, here the rolls are driven by surface-mounted PMSM, which drive the rolls without a gearbox. Taking into account the uneven load of the motors and the increased pressing force (up to 120 kN/m), the possible required increase in the rotational speed by 5%–10%, and also bearing in mind the proposed range of motors, the motors with power of 153 kW each and with $Z_p = 6$ under the rated speed rotation of 430 rpm are selected. Motor parameters are the rated voltage $V_{nom} = 337$ V, the rated current $I_{nom} = 297$ A, the rated torque $T_{nom} = 3400$ N-m, the maximum torque $T_{max} = 6120$ N-m, $\cos \varphi = 0.93$, the moment of inertia $J_m = 6.76$ kg-m², the leakage inductance $L_d = 0.28$ mH, the stator winding resistance $R_s = 0.009$ Ω, flux linkage $F_r = 1.01$ Wb.

The **Paper3b** model simulates the drive of one roll, without taking into account the elasticity of the mechanical connection; it is intended to get acquainted with the accepted control system.

The torque of PMSM with surface-mounted magnets is equal to

$$T = 1.5 Z_p F_r i_q \tag{2.15}$$

where i_q is the component of the space vector of the stator current, perpendicular to the rotor flux linkage, it is determined by the given motor torque, and in fact by the output of the speed controller. Since the i_d component does not affect the value of the motor torque, it is reasonable to take it equal to zero. Thus, the control system has to have the speed controller and controllers of the i_q, i_d currents.

The main technical problem is to determine these components; to be more exact, to determine the rotor position and its rotational speed. PMSM manufacturers usually supply them with appropriate sensors to measure these quantities, but they increase the cost of the installation and reduce its reliability. Therefore, it is advisable to be able to estimate these quantities without the use of special sensors. The **Paper3b** model provides for the possibility of using both such sensors when setting the signal *Control_mode* = 0, and estimates of the rotational speed and position of the rotor with *Control_mode* = 1. These estimates are performed as follows.

The stator flux linkage components Ψ_α and Ψ_β are computed as

$$\Psi_{\alpha(\beta)} = \frac{T_i}{sT_i + 1}\left(V_{s\alpha(\beta)} - R_s I_{s\alpha(\beta)}\right) \tag{2.16}$$

where $V_{s\alpha(\beta)}$, $I_{s\alpha(\beta)}$ are the components of the motor voltage and current space vectors; the condition $T_i \omega \gg 1$ is valid in the essential frequency range, so that the frequency characteristics of the lag elements are very close to the integrator frequency characteristics. The components of the rotor flux that are created with the permanent magnets are

$$F_{r\alpha(\beta)} = \Psi_{\alpha(\beta)} - L_d I_{s\alpha(\beta)} \tag{2.17}$$

The rotor position may be found as

$$\theta = arctg\left(\frac{F_{r\beta}}{F_{r\alpha}}\right) \tag{2.18}$$

and the rotating speed as

$$\omega(k) = \frac{\theta(k) - \theta(k-1) + 2\pi a}{Z_p T_e} \tag{2.19}$$

where T_e is the sampling period, $a = 1$, if $\theta(k)$ and $\theta(k-1)$ are taken from the different periods of rotor rotation (from the different teeth of the sawtooth waveform $\theta = f(t)$), $a = 0$ in the rest of the cases.

In order to the system would work properly, the initial estimation of the flux linkage has to be done under the start that corresponds to the real rotor position. It is carried out in the following way. During the start, after the DC link voltage has reached the steady state, the modulation voltage V_e of the small frequency, for instance, 0.5–1 Hz, and of the corresponding small amplitude is fabricated. The rotating field is created in the stator that leads the rotor; the axis of the rotor magnet flux lags the space vector V_e about $\pi/2$. In this way, calculating the position θ_e of the space vector V_e, the rotor position may be estimated. When the rotor rotates at a steady speed, the signal *Set* is formed; at that, the quantities $F_r \sin(\theta_e - \pi/2)$ and $F_r \cos(\theta_e - \pi/2)$ are set at the outputs of the lag elements; the signal from the control system that corresponds to the normal operating mode comes at the modulator input, instead of the signal V_e.

In the model under consideration, the signal θ_e is fabricated in the **Control/Start** subsystem, and the calculation of the components $F_{r\alpha}$, $F_{r\beta}$ in the **Control/Flux_Est** subsystem. For calculations according to (2.16), the blocks with first-order transfer function are used, with the possibility of setting the output by an external control signal: at $t = 2.2$ s, the *Set* signal is generated, according to which the flux linkage values fabricated to this time are rewritten to the outputs of these blocks. In the subsystems **Control/Teta_Est** and **Control/Speed_Est**, the values of the position of the rotor and the speed of its rotation are calculated according to formulas (2.18), (2.19). In blocks **Current_Reg** and **Current_Reg1**, the components of the spatial vector of the stator current i_q and i_d are controlled, respectively. Block diagram of circuits for estimation of the position and speed of rotation of the rotor is shown in Figure 2.22.

The following process is simulated: acceleration to the frequency of 0.8 Hz; storage of the values of the flux linkages fabricated by the time of $t = 2.2$ s; acceleration to a speed of 30 rad/s with an acceleration of 1.8 rad/s² at $t = 2.5$ s; the speed reference step from 30 to 32 rad/s at $t = 40$ s. The load torque is taken equal to 1.5 kN-m and the moment of inertia is 2028 kg-m².

The process with *Control_mode* = 1 is shown in Figure 2.23. It can be noted that the settling time for a reference step is approximately 1.5 s with an overshoot of 15%. The motor current is quite close to sinusoidal: THD is 4% at $t = 30$ s.

The **Paper3a** model is under consideration. Inverters of the both drives are supplied with DC voltage by the common voltage source of 500 V. This source consists of the three-phase half-controlled thyristor rectifier with the voltage controller, Figure 2.24. The rectifier output voltage U_r is

$$U_r = \frac{3 \times \sqrt{2}}{2\pi} V_{ph-ph} (1 + \cos\alpha)$$ (2.20)

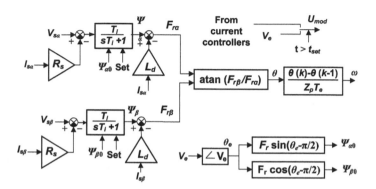

FIGURE 2.22 Block diagram of circuits for estimation of the position and speed of rotation of the rotor PMSM.

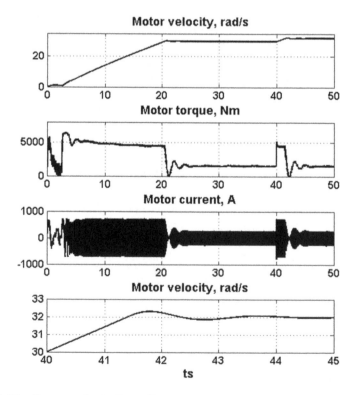

FIGURE 2.23 Processes in the **Paper3b** model.

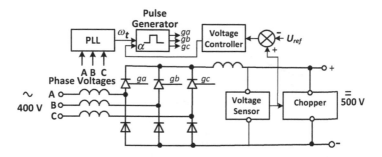

FIGURE 2.24 DC fabrication with half-controlled three-phase rectifier.

where V_{ph-ph} is the supply voltage rms value. A chopper is set at the rectifier output in order to limit the DC voltage surge under abrupt DC current change. The chopper circuits are the simplified version of the chopper from the **AC4** subsystem. It is supposed in model under consideration that the necessary sensors are installed, so that the rotor positions and their rotational speeds are known accurately.

The drive of the upper roll is the leading one, and the lower one is the slave. After the roll pressing force is established, the latter is switched to the torque control mode, and the speed controller is turned off, as implemented in the previously considered **Paper2d** model, Figure 2.14. The distribution of loads (torques) between the motors is determined by the factor K_{sh}. In the model, $K_{sh} = 0.6$ is assumed for the upper motor and 0.4 for the lower one. A circumferential force that is the same for both rolls is meant as the load T_L. The following conditions are simulated. By start, $T_L = 9$ kN; at $t = 3$ s, the motors begin to speed-up to 10% of the operating speed with the acceleration of 0.4 m/s^2. After they started to rotate, at $t = 4$ s, the load torque sharply decreased. At $t = 10$ s, the roll pressing force is set equal to half the working one (the latter corresponds to $T_L = 3.6$ kN), the distribution of loads is established, and the acceleration to the operating speed begins, which is achieved at $t = 10 + 12.7 \times 0.9/0.4 = 38.6$ s. Simultaneously, an increase in the pressing of the rolls takes place, with the rate of 60 N/s, so that the operating pressure is reached at $t = 10 + 0.5 \times 3600/60 = 40$ s, that is, almost simultaneously with reaching operating speed. At $t = 50$ s, the load increases by 20%, and at $t = 65$ s, the PM speed increases to 13 m/s.

Load torque distribution is assigned with the block **Sharing*** in the **Sharing** subsystem; the actual distribution is determined by the ratio of the motor torques at the output of the **Divide** and the **Switch** blocks. The circumferential speeds of the rolls must be equal. To fulfill this condition, a signal of sufficient magnitude acting on the load torques appears at the output of the **Saturation** block even with a small difference in speeds, which equalizes the circumferential speeds. Physically, the appearance of this effect can be explained by the fact that when the speed of rotation of one roll relative to another one attempts to increase, a tendency to slip arises, the coefficient of friction of which is much greater than the coefficient of rolling friction, and this increase in resistance prevents the speeds from divergence.

Process is shown in Figure 2.25. One can see that the wanted ratio of the loads is kept up. For example, the torques of the upper and lower motors are equal to 3.53 and 2.24 kN-m respectively at $t = 30$ s, that is, $K_{sh} = 3.53/(3.53 + 2.24) = 0.61$, and

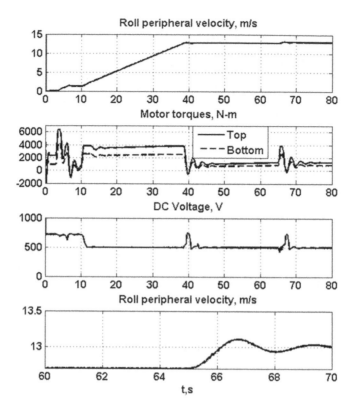

FIGURE 2.25   Processes in the drives of the press in Figure 2.19c.

at $t = 60$ s 1.26 and 0.84 kN-m respectively, that is, $K_{sh} = 1.26/(1.26 + 0.8) = 0.61$. If to repeat the simulation without chopper, the values of the DC voltage surges will be much larger.

The **Paper3a2** model is the same as the **Paper3a** model, but instead of the sensors, the estimated values of the rotational speeds and rotor positions are used, as it was described above. At first, the voltage with the frequency of 0.8 Hz and the relative amplitude of 0.03 is applied to the motors, they are speeding up. At $t = 2.2$ s, the flux linkage values that have been calculated to this instant of time is sent to the blocks for the flux linkage computations (the **Flux_Est** subsystems) where they are considered as the initial values; subsequently, the outputs of these blocks are used to calculate the rotor positions and rotational speeds. The speed reference that equals 10% of the operation one is given at $t = 2.5$ s, and, in what follows, process proceeds just like in the previous model. Process is shown in Figure 2.26. It can be seen that it almost does not differ from that shown in Figure 2.25.

The press whose structure is shown in Figure 2.19d is simulated in the **Paper3c** model. The upper and lower rolls are driven; the lower roll drive is the leading one, since it has a larger wrap angle. The diameter of the lower roll is assumed to be 0.85 m with a moment of inertia of 2300 kg-m², and the upper one is 0.8 m with a moment of inertia of 1800 kg-m². The diameter of the granite roll is 0.95 m with a

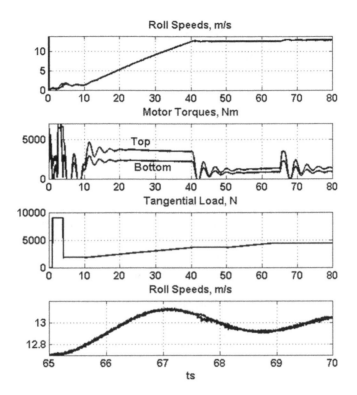

FIGURE 2.26 Processes in the model **Paper3a2**.

moment of inertia of 3000 kg-m². The total load is increased by 20% compared to previous **Paper3#** models. For unification, the same motor as in the **Paper3#** models is used for the both rolls.

The **Paper3a** model was taken as a basis when **Paper3c** model was created, with adequate parameter change. The main change is consideration of the existence of the third, granite, non-driven roll. As mentioned above, since the circumferential speeds of the rolls are equal, when the speed of a roll is trying to increase, there is a tendency for mutual sliding (instead of rolling), at which the moment of resistance increases sharply, which keeps the rolls in synchronous rotation. This phenomenon is used to simulate the non-drive roll in accordance with the diagram shown in Figure 2.27.

Here M_{st} is the torque of resistance when the roll rotates that is determined mainly by the bearing friction, M_g is the torque which is demanded for the roll rotation and acceleration. This torque is defined as the difference between the circumferential speeds of the driving bottom roll and the granite roll; this difference is amplified afterward with the high gain K and filtered. The corresponding circumferential force $F_g = M_g/R_g$ (R_g is the radius of the granite roll) is added to the predetermined circumferential force values to form the total force that the drive rolls must fabricate.

It is supposed in the models that at $t = 6s$, when the driven rolls rotate with the speed of 10% operating, they are pressed to the granite roll, which, hereupon, speeds up, so that all three rolls are rotating synchronous. At $t = 10s$, the press speeding up to the

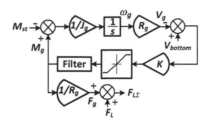

FIGURE 2.27   Simulation of the non-driven roll.

operating speed begins, as in the previous models. Process is shown in Figure 2.28. It can be seen that the circumferential speeds of the rolls are equal, and the specified load ratio is maintained. For example, at $t = 30$ s, the lower (leading) roll and upper one torques are 3.8 and 2.5 kN-m, respectively; $K_{sh} = 3.8/(3.8+2.5) = 0.6$, and at $t = 60$ s 1.6 and 1.03 kN-m, respectively, that is, $K_{sh} = 1.6/(1.6+1.03) = 0.61$. It is seen also as a significant influence of a heavy granite roll on the dynamic loads of the driven rolls.

The **Paper3c2** model is the same as the **Paper3** model, but, instead of the measured values of the motor rotor positions and rotational speeds, the estimated quantities are used as in the **Paper3a2** model. The processes in them almost do not differ from those observed in the **Paper3c** model.

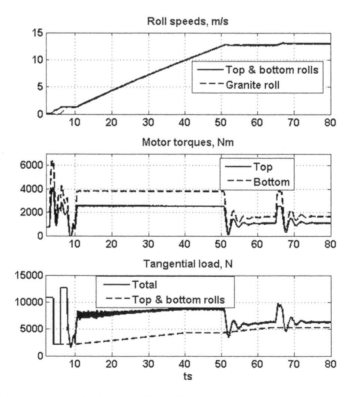

FIGURE 2.28   Processes in the model **Paper3c**.

2.2.3 DRYER SECTION

After the press section, the dryness of the paper is typically between 28% and 45%. Further dehydration to a final dryness of 92%–95% occurs in the dryer section of PM. The main method of paper drying is the contact method with the help of drying cylinders (another term—drying can) heated by steam, to which the paper is pressed with the help of felts. In modern PM, the diameter of the cylinders reaches 1.8 m (1.5 m in old designs). The number of cylinders is 40–60 or more. At the end of the dryer section, to cool the paper web before it enters the calender, 1–2 cooling cylinders are installed, and cooled by running water. Since shrinkage of the paper occurs during the drying process, the circumferential speeds of the cylinders must be able to vary. For this purpose, the dryer section is divided into several groups, between which the ratio of speeds can change. The number of groups is usually 5–6 and can reach 10, and the number of cans in a group is 6–12 and sometimes up to 20, depending on the type of product.

Most often, the cylinders in the group are placed in two rows in a staggered order, so that when passing through the group, the paper is pressed against the cylinders alternately by one side or the other, Figure 2.29; there are other cylinder arrangements: in one and in three rows. In most PMs, a group electric drive of cylinders is used, in which one motor transmits rotation with the same speed to all cylinders in the group through a system of gear wheels. In principle, there are other systems: with an individual drive of each cylinder and with non-driven cylinders that rotate due to the efforts of the felts pressed against them. Only the group drive is considered below, when all the cylinders in the group form one mechanical system with the total moment of inertia J_L, having the rotational speed ω_L and the peripheral speed V_L; this mechanical system is driven by the motor with the help of elastic shaft. To simplify the system, the refrigeration cylinders mentioned above are included in the last drying group.

The drying group is considered to consist of 12 drying cans with a diameter of 1.8 m. Paper width is 4.8 m and the PM speed is 762 m/min. In accordance with the method of specific powers, it can be concluded from [2], assuming that with an increase in the diameter of the cylinder, the required power increases proportionally:

$$P_{NRL} = 0.0018 \times 96.4 \times \frac{1.8}{1.5} \times 4.8 \times 7.62 \times 12 = 91 \text{ kW}.$$

Taking $P_{RDC} = 2P_{NRL}$, the motor power is 182 kW. Note that the found values of P_{NRL} and P_{RDC} correspond to the powers given in [3] for a drying group of 12 cylinders with a diameter of 1.5 m (83 and 179 kW, respectively). However, when choosing motor power,

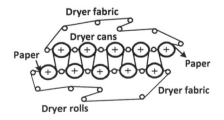

FIGURE 2.29 Arrangement of the drying cans in the group.

two more points must be taken into account: the need to accelerate the load with a large moment of inertia and the presence of condensate in the cylinders during start-up.

With the peripheral speed of 12.7 m/s, the cylinder rotational speed is $12.7/0.9 = 14.1$ rad/s $= 134.8$ rpm. IM with $Z_p = 4$ is chosen that drives the group through the gearbox with $i = 4.8$.

A similitude method is used to compute the total moment of inertia. The value of the moment of inertia of the drying group that consists of 12 cans with a diameter of 1.5 m at the same width of the paper web which is equal to 76,005 kg-m² is given in [3]. Since the moment of inertia is proportional to the fourth power of the diameter (with a proportional increase in wall thickness and the same material), we have $J_L = 76,005 \times (1.8/1/5)^4 = 157,600$ kg-m². Taking into account that the increase in wall thickness is not mandatory, and the increase in the moments of inertia of the gears can be less, let's take $J_L = 130,000$ kg-m², its value, reduced to the motor shaft, $J_{Lr} = 5642$ kg-m².

The effect of the presence of a condensate in the cans on the value of the load torque is considered in [13–15]. These works provide data on the increase in torque depending on the thickness of the condensate layer or the condensate volume content and on the speed of the PM for a cylinder with a diameter of 1.8 m. At start-up, the torque quickly increases to M_{cm}, and then sharply decreases to a small value. An approximate approximation of the dependences given there is shown in Figure 2.30 for the dryer section under consideration (12 cylinders with a diameter of 1.8 m and a paper web width of 4.8 m) and with a relatively small amount of condensate content (the thickness of the condensate rim is 1.6, 3.2, and 6.4 mm, which corresponds to approximately a volume content of 0.44%, 0.88% and 1.74% respectively). Such condensate content is considered normal if the cylinders are equipped with additional elements (Spoiler bars) [14,15]. From these data, it follows that even with a small amount of condensate, the load torque increases significantly.

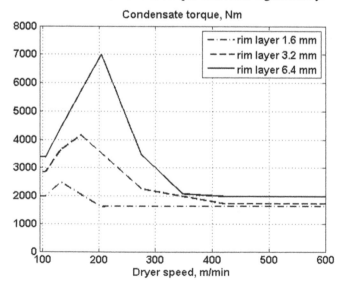

FIGURE 2.30 Dependence of the dry section torque on the condensate content and PM speed.

If to take speeding-up time equal to 240 s [14,15], the demanded dynamic torque is $M_{din} = 5642 \times 14.1 \times 4.8/240 = 1591$ N-m. The normal load that corresponds to power of 91 kW is $M_{NRL} = 1345$ N-m. During 2 minutes, the motor can carry a load equal to 1.5 of the rate; so, assuming the process to be adiabatic, the load should not exceed $\sqrt{1.5^2 \times 2/4} = 1.06$ for 4 min. Considering this circumstance, as well as the fact that the actual value of the moment of inertia, taking into account the moment of inertia of the motor, may exceed the value taking for calculation, that in some cylinders the remaining amount of condensate may exceed 1% [15] and that, because of oversight, the doctor blade can be in the working position at start, the motor with power of 250 kW is chosen, with a rated torque of 3236 N-m. IM has the rotational speed of 738 rpm, the voltage of 690 V, the rated current 266 A, the non-load current 99 A, the maximal torque M_m, the starting torque and the stating current are equal to 1.9, 0.8 and 4.5 pu, respectively, the moment of inertia 13.1 kg-m², reduced to the load $J_m = 302$ kg-m², the efficiency 0.947, cos $\varphi = 0.83$.

The stiffness coefficient C is computed by (1.15), (1.16) with $D = 0.125$ m, $l = 3$ m [5], $\tau = 80$ N/mm²,

$$C = \frac{\pi \times 80 \times 0.125^4 \times 10^9}{32 \times 3} = 0.63 \times 10^6 \text{ N-m/rad,}$$

that is, $\omega_r = \sqrt{\dfrac{0.63 \times 10^6 \times (130,000 + 302)}{130,000 \times 302}} = 45.7$ rad/s.

The damping coefficient B is computed by (2.4) as $B = 0.07 \times 2 \times 0.63 \times \dfrac{10^6}{45.7} = 1378$ N-m-s/rad.

DTC system is used for the motor control, the **Paper4** model; such a system was used before in the **Paper2** model. The torque produced by the presence of the condensate in the cans is added to the load static torque that is equal to $1345 \times 4.8 = 6.46$ kN-m; it is supposed that the condensate content is 0.88% (thickness 3.2 mm). The condensate torque is computed with the help of the **Table** block, for which the input quantity is the PM linear speed and the output corresponds to Figure 2.30. This model gives an opportunity to investigate a number of the dryer section operation modes.

The steady-state is investigated at first. Since reaching such a mode demands for a long simulation, the **Paper4_1** model is made, in which the special measures are applied to accelerate the achievement of a steady state, which is already achieved at $t = 100$ s. Figures 2.31 and 2.32 show the system's response to a 5% speed reference step change with and without a notch filter. Speed controller gains are taken as $K_p = 170$, $K_i = 60$ in Figure 2.31. It is seen that the notch filter does not affect processes in the system. But if to take these gains half as much again, the system without the notch filter is instable, whereas with this filter, process in the system is quite satisfied, Figure 2.32. Therefore, taking in mind possible deviations of the system parameters from rated, further simulations were performed with the filter. Figure 2.33 shows the response of the system to an increase in load torque by 13%. The maximum speed deviation is 0.15%. In these experiments, $Mode = 0$. This model makes it possible to simulate the effect of backlash at $Mode = 1$. In this case, in the interval 100 s $< t < 120$ s, the load torque drops to zero, for example, for the last dryer section, due

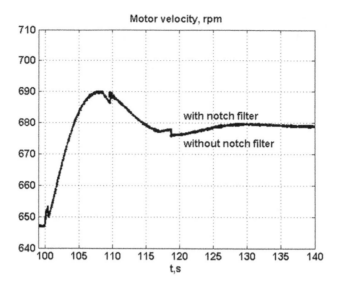

FIGURE 2.31   Dryer section step response with and without the notch filter, with the accepted controller gains.

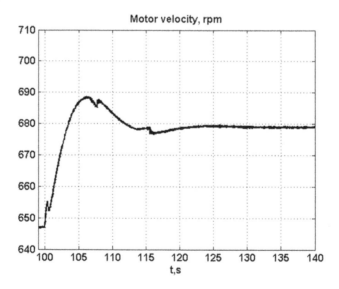

FIGURE 2.32   Dryer section step response with notch filter under gain increase by 1.5 times.

to the influence of the calender [3,16]. The simulation result with a backlash value of 2° is shown in Figure 2.34. It can be seen that this phenomenon leads to a short-term change in the signs of the motor torque and elastic torque, the values of which, during their subsequent recovery, exceed the previously observed steady-state values.

Let's consider the **Paper4** model. The following process is simulated: At first, the motor speeds up to the speed of 10% operating one with the static load of (1200 + 1200)

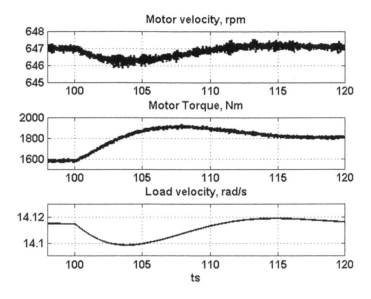

FIGURE 2.33 Dryer section response when the load surge.

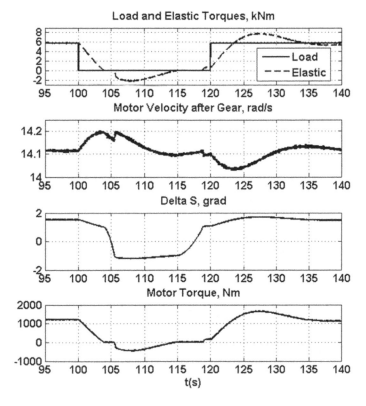

FIGURE 2.34 Dryer section, effect of backlash of 2°.

N-m where the second term is the breakaway torque. After the beginning of the rotation, this term disappears, but the additional load torque appears because of the condensate presence. At $t = 20$ s, speeding up to the operating speed begins, with the
acceleration of 2.7 rpm/s. When the operation speed is reached, at $t = 270$ s, the speed
reference increases by 5%, and at $t = 360$ s the load torque increases by 20%. The process is shown in Figure 2.35; it is seen that the real speed corresponds to the wanted
value. The motor torque at the start first instant increases to 4.8 kN-m, it decreases after
rotation beginning, then, after passing the value of 3.6 kN-m, with the speed rise and
the condensate reducing, decreases to 3.2 kN-m and, after speeding up end, decreases
to 2 kN-m; this, with the rated torque of 3.2 kN-m, may be considered as acceptable.

The **Paper4a** model simulates the steady state condition of the last two dryer
sections connected by a paper web. The sections are taken the same, the models of
which are taken from the **Paper4_1** model, and the web tension model is taken from
the **Paper1d1** model. The change in tensile strain is described by equation (2.6), the
tension $T = QE\varepsilon_2$, where, as before, $Q = 245 \times 10^{-6}$ m^2, $E = 0.8 \times 10^{10}$ N/m^2. We will
consider ε_1 as a degree of "shrinkage" of the paper web and take $\varepsilon_1 = 0.005$. Since it
follows from (2.6) that in order to obtain a given tension T^*, the ratio of the circumferential speeds of the rolls must be $v_2/v_1 = (1 + T^*/QE)/(1 + \varepsilon_1)$, then, after calculations for $T^* = 1500$ N, it will be $v_2/v_1 = 0.9958$, or, with a given rotational speed of
the last but one section motor of 647.1 rpm, the rotation speed of the motor of the last
section should be equal to 644.4 rpm.

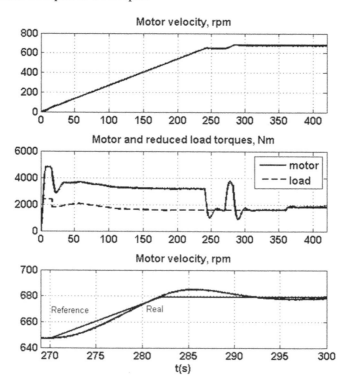

FIGURE 2.35   Process in the **Paper4** model.

It is assumed in the model under consideration that a steady state is reached at $t = 100$ s and the tension calculation unit is activated; at $t = 200$ s, the calculated tension value of 1500 N is set. At this moment, the load of the last section increases by 20%, which leads to a short-term drop in the motor speed and, accordingly, to a decrease of the tension between the sections. After the motor speed is restored, the tension increases and as a result reaches the original value. The process is shown in Figure 2.36. As it can be seen, even a slight decrease in the rotational speed (maximum speed drop of 0.07%) causes a noticeable change in tension, and the natural process of its recovery is extremely slow, so that in critical cases the use of a tension controller is required. It should be noted, however, that in reality the changes of load in the PM do not proceed abruptly, as in this model, but more smoothly, so that the change in tension will be less.

2.2.4 CALENDER

The calender is installed after the dryer sections and is designed to increase the gloss, smoothness, and bulk density of the paper and give it a uniform thickness along the web width. The calender consists of several rolls arranged vertically and pressed against each other which the paper web sequentially bends around. One of the rolls, usually the lower or second from the bottom is driven, and the rest rotate due to friction against each other. In previous PM designs, the number of rolls was 6, and sometimes more, but in new designs, the rolls are equipped with additional devices that make it possible to obtain very high values of linear pressure (up to 200 kN/m) with a smaller number of rolls, so that the number of rolls is 2–4. The rolls are mostly solid cast iron. Figure 2.37 shows a four-roll calender with a lower roll drive.

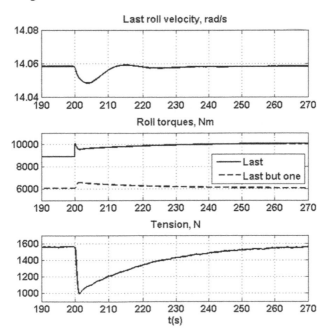

FIGURE 2.36 Processes in the **Paper4a** model.

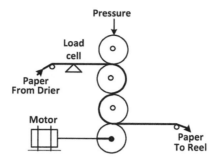

FIGURE 2.37 Four-roll calender with a lower roll drive.

The calender electric drive control system must ensure stable operation both with a relatively small moment of inertia of one roll before pressure is applied and with a full moment of inertia of four rolls after pressure application. In this case, in the first mode, the rotational speed is controlled, and in the second mode, the tension is controlled, the specified value of which for the considered PM (speed 12.7 m/s, web width 4.8 m) is taken equal to 1200 N. The roll diameters can be either the same or somewhat different. In [10], an example of a calender with four identical rolls of 0.61 m in diameter, operating at the pressure of about 80 kN/m is given.

The calender of the same design but working with the pressure of 200 kN/m is considered. The rotational speed of the rolls is $12.7 \times 30/0.305/\pi = 398$ rpm. Assuming that the circumferential forces are proportional to the pressure and taking into account that for the calender in [10] the motor power is assumed to be 117 kW, the induction motor with a power of 315 kW at a voltage of 690 V, 1490 rpm is chosen, which rotates the calender through the reducer $i = 3.2$. Motor parameters: the rated current 320 A, the non-load current 110 A, the maximal torque M_m, the starting torque, and the stating current are equal to 2.1, 0.7, and 5.6 pu, respectively, the moment of inertia 4.7 kg-m², reduced to the load $J_m = 48.1$ kg-m², the efficiency 0.962, cos $\varphi = 0.86$. It can be noted as an alternative the use of IM with $Z_p = 6$ without a gearbox, but such motors of the same series (AXR manufactured by ABB) have twice as much weight (3870 kg instead of 1900 kg) and consume much more current (412 A); of course, other options are possible too.

The stiffness coefficient is taken the same as in Section 2.2.3: $C = 0.63 \times 10^6$ N-m/rad. Since the moment of inertia of such a calender is determined as 1910 kg-m², then

$$\omega_r = \sqrt{\frac{0.63 \times 10^6 \times (1910 + 48.1)}{1910 \times 48.1}} = 115.9 \text{ rad/s}.$$

The damping coefficient B is calculated with using (2.4) as $B = 0.07 \times 2 \times 0.63 \times \dfrac{10^6}{115.9} = 761$ N-m-s/rad.

DTC system is used, as in the **Paper2** model, for motor control (the **Paper5** model). The difference is that, in the latter model, the opportunity to change the moment of inertia and the load torque is provided by the signal *Press*. It is taken that the motor operating load is equal to 50% of the rated load, that is, 1 kN-m.

The process is shown in Figure 2.38 when the calender speeds up to the speed of 10% operating one, the load torque and the moment of inertia are equal to ¼ of

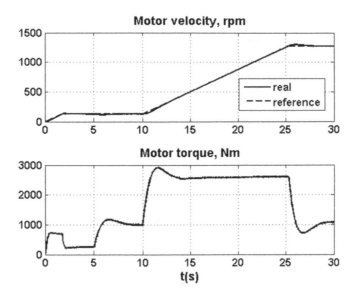

FIGURE 2.38   Calender speeding up.

the operating values. The pressure is applied at $t = 5$ s, all four rolls turn out to be in contact as a single unit, and the load torque rises. At $t = 10$ s, the calender speeding up to the operating speed begins, with the motor torque of 2600 N-m that exceeds the motor rated torque by 1.3 times but is permissible, taking into account a short time of speeding that ends at $t = 25$ s.

Figure 2.39a and b show the process without and with the notch filter, with a step change in the speed reference of 40 rpm (3.1%). It can be seen that when the filter is applied, the oscillation is significantly reduced. Moreover, even with a small increase of the controller gains, the system without filter is oscillating, whereas with the notch filter, it remains stable, so further studies are carried out with this filter. The settling time is 1 s. Figure 2.39c and d shows the process with a stepwise increase in load torque by 20%. It can be seen that due to the large moment of inertia of the load, the motor torque increases smoothly, within 1 s, the maximum drop of the rotational speed is 0.4%.

The **Paper5a** model simulates the joint operation of the calender with the last dryer section, the model of which is taken from the **Paper4a** model; the tension fabrication model is also borrowed from there. The back tension for the dryer section and the forward tension for the calender are assumed to be constant and are assumed to be taken into account in the values of the load torques TL. The value of ε_1 is taken equal to zero, and the tension between the calender and the dryer section is 1500 N. Then the ratio of circumferential speeds is $v_2/v_1 = 1 + \dfrac{1500}{245^{-6} \times 0.8 \times 10^{10}} = 1.000765$.

It is assumed in the simulation that by the time $t = 150$ s, a steady state is reached, and the tension controller is turned on which affects the rotation speed of the calender motor. Figure 2.40 shows the process when, at $t = 180$ s, the tension reference increases from 1.5 to 1.8 kN.

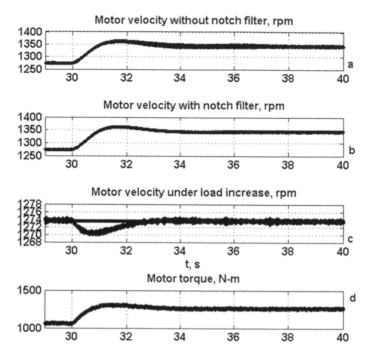

FIGURE 2.39 Calender drive response to stepwise changing in the speed reference and the load torque (a) speed reference change without the notch filter; (b) speed reference change with the notch filter (c) speed change under load torque changing; (d) motor torque change under the load torque changing.

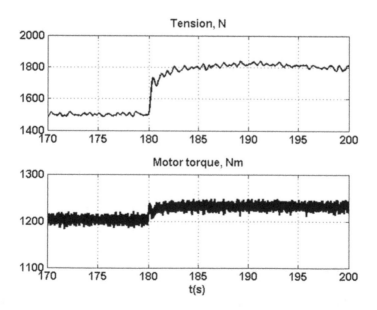

FIGURE 2.40 Processes in the **Paper5a** model when the tension reference increases.

The tension changes when the modulus of elasticity E increases from 0.8×10^{10} to 1×10^{10} for 5 s, beginning from $t = 180$ s, without tension controller (i.e. with the invariable speed ratio) and with such controller is shown in Figure 2.41. It is seen that in the first case, the tension increases significantly, while in the second it remains unchanged.

The processes of changing the load of the calender are shown in Figure 2.42. At $t = 150$ s, the load torque increases by 20% for 5 s with the tension controller off. Although the reduction in the speed of rotation of the calender is small, due to the large value of the coefficient of elasticity, this leads to a significant decrease in the tension between the calender and the last dryer section which is slowly restored within 50 s. At $t = 200$ s, the tension controller is turned on, and at $t = 230$ s, the load torque returns to its original value. At that, the increase of the tension is much less and quickly, within 6 s, returns to the set value.

2.2.5 REEL

The reel (winder, coiler) is set in the PM end and is intended for winding paper into rolls. One distinguishes between the axial and peripheral reel. The first one is the same in principle as the sheet metal winders discussed in the first chapter. Now such reels are practically not used. In the peripheral reel, the paper roll that is wound is pressed against the reel drum which is rotated by an electric motor with the constant circumferential and angular speeds. The main advantage of such a reel is an ability to carry out paper production without stopping after the end of the winding of the running roll. This is done as follows. The device has several reel spools for paper winding. While paper is being wound on one of them, the second is on the receiving arms. When the winding of the running roll ends, the paper starts to wind on this second spool. At this time, the already wound roll, which is in the main arms, is removed from the reel, and the wound roll is transferred from the receiving arms to the released main arms. The paper roll rotates under the action of the peripheral force between the

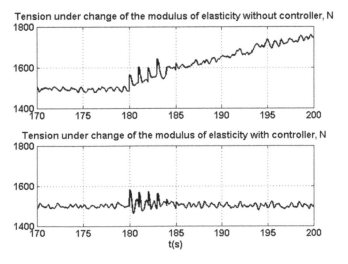

FIGURE 2.41 Processes in the **Paper5a** model when the modulus of elasticity increases.

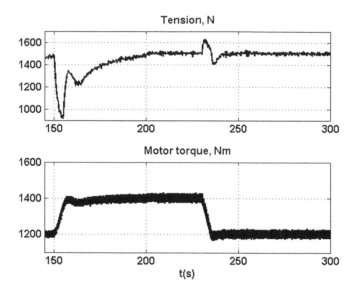

FIGURE 2.42 Processes in the **Paper5a** model when the load changes.

surfaces of the roll and the reel drum. As the diameter of the winding roll increases, its rotational speed continuously decreases, and the circumferential speed remains constant, equal to the circumferential speed of the drum. When the roll diameter rises, the distance between the axis of the drum and spool increases by the turn of the main arms; the pneumatic cylinders are used to ensure a constant force of pressing the drum to the roll. The layout of the peripheral reel is sketched in Figure 2.43.

It will be modeled the same PM as earlier (the speed $V = 762$ m/min $= 12.7$ m/s, the web width $b = 4.8$ m). The thickness of the paper $h = 51$ mkm with a mass of 60 g/m², so that the paper density $\gamma_p = 60$ g/m²/51 mkm $= 1176$ kg/m³.

The drum diameter is taken as $D_d = 1.1$ m, with the mass $m_c = 6248$ kg and the moment of inertia $J_d = 1790$ kg-m², the reel spool diameter $D_s = 0.42$ m, its moment of inertia $J_s = 60$ kg-m², the roll maximal diameter $D_{rm} = 2.54$ m. The duration of the winding of one roll can be found in the obvious relationship

$$t_w = \frac{0.25\pi\left(D_{rm}^2 - D_s^2\right)}{VQ} = \frac{\pi\left(2.54^2 - 0.42^2\right)}{4 \times 762 \times 51 \times 10^{-6}} = 126.7 \text{ min.}$$

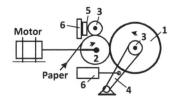

FIGURE 2.43 Layout of the peripheral reel (1-paper roll, 2-reel drum, 3-spools, 4-main arms, 5-receiving arms, 6-pneumatic cylinders).

The mass of the full paper roll, with a density of 60 g/m^2, is $m_r = 126.7 \times 762 \times 4.8 \times 0.06 = 27.8$ t. The drum rotational speed is $12.7 \times 30/0.55/\pi = 220.6$ rpm. According to [2], the drum drive power, not taking into account the power of tension, is $P_1 = 0.008 \times 96.4 \times 4.8 \times 7.62 = 28.2$ kW. Assume that the paper has a tensile strength of 5 kN/m and a tension is equal to 0.083 of the tensile strength, that is, is equal to $T_r = 5 \times 4.8 \times 0.083 = 2$ kN, then the tension power is $P_2 = 2 \times 12.7 = 25.4$ kW. Considering the available and described in the literature PM, one can see that the NRL power is approximately twice the tension power, and the installed power is approximately by the same times more than NRL power [17]. Taking this in mind, the PMSM with $Z_p = 6$ is selected which operates with the frequency of 30 Hz at the voltage of 336 V. The rated power is 107 kW, the rated torque is 3400 N-m, and the rated current is 210 A, $\cos \varphi = 0.93$, $J_m = 6.76$ kg-m^2.

The motor torque M_m is used for creating tension T_r, for overcoming the friction torque in the drum bearings M_{fd} and for rotating the paper roll: $M_m = T_r R_d + M_{fd} + F_{dr} R_d$ where F_{dr} is the tangential contact force between reel drum and paper roll. In turn, this value is determined by the rolling friction force between the drum and the roll F_{fr}, the friction force in the reel spool bearings F_{fs}, the friction force of the doctor on the drum F_{scr}, and also by the torque M_{din0} required to increase the kinetic energy of the roll when winding with a constant speed. When the motor rotational speed changes, the dynamic torque M_{din} appears; when it is calculated, the drum and the roll are considered as a single system with a common inertial mass.

Let's calculate the above values, assuming the force of pressing the roll to the drum $F_p = 20$ kN and using the technique from [18]. The value $M_{fd} = f_c Q_c\, d_d/2 = 0.003 \times 65.6 \times 0.4/2 = 0.039$ kN-m where Q_c is the resultant of a gravitational force of the drum and a pressing force (it is supposed that the axes of drum and reel spool are on the same horizontal), $Q_c = \sqrt{62.5^2 + 20^2} = 65.6$ kN, d_d is a neck diameter, f_c is a friction coefficient. The value $F_{fs} = f_c Q_s d_s/D_r$ where Q_s is the resultant of a gravitational force of the roll and the pressing force, d_s is the diameter of the reel spool neck (0.4 m). The value Q_s depends on the roll diameter,

$$Q_s = \sqrt{(gm_r)^2 + F_p^2},\qquad (2.21)$$

где g is the acceleration of gravity. Approximately assuming that the mass of the hollow reel spool is equal to the mass of an equal-sized cylinder filled with paper, it follows $m_r = 0.25\pi b \gamma_p D_r^2 = 4431 D_r^2$ kg, so that m_r value changes from 781 to 28,588 kg, and F_{fs} value from 61.2 to 133 N. The friction force of the doctor on the drum $F_{scr} = f_{scr} F_{pscr} b$ where $f_{scr} = 0.18$ is the doctor friction coefficient, $F_{pscr} = 150$ N/m is a linear pressure between the doctor and the drum, then $F_{scr} = 0.18 \times 150 \times 4.8 = 130$ N. The rolling friction force between the drum and the roll depends on the roll radius:

$$F_{fr} = 2f_{fr} F_p (1/D_d + 1/D_r),\qquad (2.22)$$

where f_{fr} is the friction coefficient equal to 0.0022 m, from which it follows that, during winding, F_{fr} changes from $F_{frmax} = 0.0044 \times 20 \times (1/1.1 + 1/0.42) = 290$ N to $F_{frmin} = 0.0044 \times 20 \times (1/1.1 + 1/2.54) = 115$ N.

The dynamic torque demanding for changing the roll kinetic energy is

$$M_{din0} = \frac{d}{dt}(J_{rs}\omega_r) \tag{2.23}$$

where $J_{rs} = J_s + J_r$, J_r is the moment of inertia of the paper roll

$$J_r = \pi\gamma_p b(D_r^4 - D_s^4)/32. \tag{2.24}$$

When the drum rotates with the constant speed (V is constant), $\omega_r = V/R_r$ and

$$M_{din0} = V\frac{d}{dR_r}\left(\frac{AR_r^4 + B}{R_r}\right)\frac{dR_r}{dt} \tag{2.25}$$

where $A = 0.5\pi\gamma_p b$, $B = J_s - 0.5\pi\gamma_p bR_s^4$,

$$\frac{dR_r}{dt} = \frac{h}{2\pi}\frac{V}{R_r}. \tag{2.26}$$

After manipulation, the peripheral force is

$$F_{din0} = \frac{M_{din0}}{R_r} = \frac{V^2 h}{2\pi}\left(3A - \frac{B}{R_r^4}\right) \tag{2.27}$$

Since $A = 8862$ kg/m², $B = 44$ kg-m², then $F_{din0min} = 5.2$ N and $F_{din0max} = 35$ N. These values can be neglected usually.

Of course, the calculated values are approximate and, when modeling a specific PM, should be refined based on the results of testing an existing PM [19] or its proto-type, or after studying the design documentation for a new PM.

The **Paper6a** model models the reel taking into attention the above given rela-tionships. The electric drive control system is the same as in the **Paper3a** model. The coefficient C, which describes the elastic connection, is preserved, since the transmitted torques are approximately the same. The load torque is equal to the sum of the tension torque, which is assumed to be constant, and the sum of the torques of the losses calculated in the **Add_Load** subsystem. The resonant frequency depends little on the roll radius and can be estimated as

$$\omega_r = \sqrt{\frac{3.2\times10^5\times(1850+6.76)}{1850\times6.76}} = 218 \text{ rad/s.}$$

A notch filter is used to reduce the oscillation. In the **Load** subsystem, the roll radius is calculated according to (2.26). To speed up the simulation, the thickness of the paper being wound is significantly increased in some cases. Also, to speed up the simulation, the inverter is powered by DC voltage. The model is intended to deter-mine the main properties of the reel electric drive.

As the main characteristic of the electric drive control system, the response to a stepwise increase of the speed reference by 10% is considered. It is obvious that the transient process essentially depends on the moment of inertia of the system, that is, from the radius of the roll.

Figure 2.44 shows the change in the rotational speed of the motor for three values of the radius: 1.27, 1, and 0.21 m (no roll). The speed controller gains were equal to $K_p = 9600/K$, $K_i = 8000/K$ where the values of K for these three cases were equal to 1, 2, and 15, respectively, that is, they changed approximately in inverse proportion to the value of the total moment of inertia. To on-line estimate this quantity, the value of the winding roll radius and the paper parameters must be known. One possibility is to use the position sensor of the arm that presses the roll against the drum [17]. Since it is sufficient to have a rough estimate of the roll radius to correct the gains of the speed controller, and the number of correction points is also small, it can be proposed to fix the number of motor revolutions after the start of winding the next roll and, after multiplying by the nominal paper thickness, obtain an approximate estimate of the radius.

In the **Paper6b** model, a subsystem for tension calculation **Tension**, taken from the **Paper5a** model, has been added to the **Paper6a** model. The simulation is performed in a simplified way: it is assumed that the speed of the calender before the reel is constant. When the tension between the calender and the reel is 2000 N, the circumferential speed ratio is $v_2/v_1 = 1 + \dfrac{2000}{245^{-6} \times 0.8 \times 10^{10}} = 1.001$, that is, at the calender speed of 12.7 m/s, the reel drum rotational speed should be equal to 12.713 m/s.

In the Calender-Reel span, the tension control is switched to speed control during the change of the shipping roll. Immediately after the paper sheet is transferred to the new empty reel spool, the reel drive is switched back to tension control mode [17]. In the **Paper6b** model, tension control is performed indirectly by setting the motor torque, taking into account losses and dynamic torque requirements. The transition from the speed control mode to the tension control is performed by the *Mode* signal

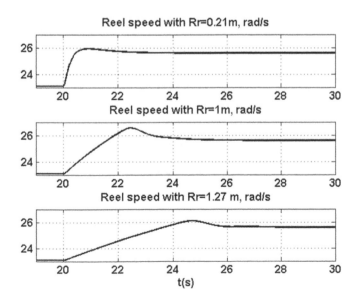

FIGURE 2.44 Response dependence on the roll radius for the reel drive.

in the **Drive** subsystem. The fabrication of the speed and torque reference values is performed in the **Control** subsystem, as shown in Figure 2.45.

When working in the tension control mode, and at this time the paper web breaks, the motor speed begins to increase rapidly and may reach unacceptable values. To prevent this effect, the **Control** subsystem provides for reducing the torque reference down to zero if the rotation speed exceeds the specified value by a certain amount (assumed to be 2.5%). The **Tension** subsystem provides the ability to simulate a web break by generating a *Break* signal.

The **Paper6b** model lets to make a number of investigations.

Figure 2.46 shows the response to a stepwise change in the tension reference. The first plot assumes that by the time $t = 10$s the reel drum is rotating with the required speed and the paper roll is being griped up. The system operates in speed

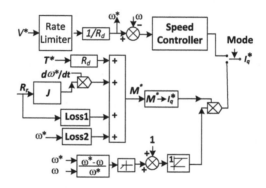

FIGURE 2.45 Block diagram of fabrication of the motor current reference.

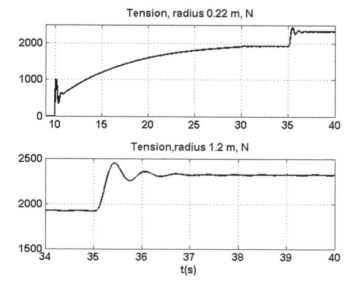

FIGURE 2.46 The reel response to a stepwise change in the tension reference under different roll radius.

control mode. According to the selected speed reference, the tension increases to the wanted value of 2 kN. At $t = 30$ s, the electric drive switches to the torque (tension) control mode. At $t = 35$ s, the tension reference increases to 2.4 kN. It can be seen that the response for a new set point occurs quite accurately and quickly. In the second plot, the winding process approaches completion with the setting of 2 kN with an error of 4%. Here, also, the response for a new set point occurs quite accurately and quickly. Of course, such a high accuracy of tension control is associated with the assumption of accurate knowledge of the loss components, which is not always true.

Figure 2.47 shows the process when, starting from $t = 35$ s, the process of the torque control reaches the steady state, the modulus of elasticity increases in 5 s from 0.8×10^{10} N/m^2 to 10^{10} N/m^2, for example, due to a change in humidity. It can be seen that, despite the rather rapid change in the properties of the paper web, the specified tension is kept.

The process of the PM speed increase is simulated in the **Paper6b1** model; the speed increases for 5 s by 2%, beginning at $t = 35$ s. The drum electric drive operates in the torque control mode and speeds up by the torque rise in proportion to the total acceleration. The process is shown in Figure 2.48. It can be seen that the wanted tension value is maintained. It is easy to see that if to turn off the *ACC* input in the **Control** subsystem, then during acceleration the tension will decrease significantly.

Let's simulate, with **Paper6b** model, the paper web break that happens at $t = 35$ s. It can be seen in Figure 2.49 that the tension goes down to zero at this instant, and the motor begins to speed up; but, when the circuits for the torque limitation are switched on, the speed rise is limited and not large.

In the **Paper6** model, the operation of the reel and the calender together is simulated. Models of electric drives are taken from the respective models described earlier. The **Tension** subsystem is taken from the **Paper6b** model with the difference that the

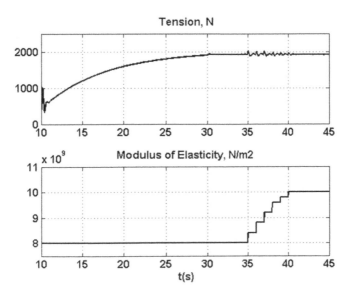

FIGURE 2.47   Process in the reel under change of the modulus of elasticity.

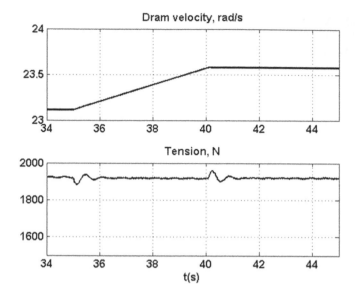

FIGURE 2.48   Tension keeping in the **Paper6b1** model.

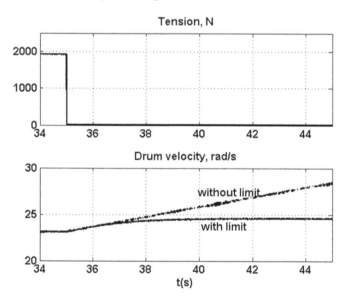

FIGURE 2.49   Speed limitation when the paper web breaks.

speed V_1 is not constant, but is determined in the calender model. The model uses the developed model of the rate limiter block which permits to change rate of output alternation during the simulation and, at the same time, unlike the library model, allows simulation with a variable-step solver. A high rate, which is unrealistic in practice, is used at the beginning of the simulation to quickly achieve an operating speed, and a lower, real rate is used when studying processes in the system. The model makes it possible to investigate various processes in the calender-reel span. For example, the process of the PM speed increase by 2.5%, beginning from $t = 45$ s, is shown in Figures 2.50 and 2.51; at $t = 25$ s, the winding of the new roll in the speed control mode begins, the tension rises smoothly, at $t = 40$ s the torque control mode is switch on. To receive the value of acceleration, the PM speed reference is used, which contains fewer higher harmonics than the rotational speed signal of the calender motor, but the following problem arises: since the rotational speed of the calender changes slowly, when the speed reference increases, its rotational speed lags behind the set value, and the reel drive reacts to this change without delay, which leads to a significant increase in tension at the beginning of the transient. To reduce this effect, a time-lag block is installed before the differentiation block in the reel control subsystem, simulating this delay. The value of the tension deviation depends on the rate of change of the reference signal. In Figure 2.50, the speed up rate is 12.7/120 m/s², and in Figure 2.51 12.7/60 m/s². It can be seen that the tension deviation is larger in the second case.

Process when the tension reference increases by 20% beginning from $t = 45$ s is shown in Figure 2.52. It can be seen that it causes a short deviation of the calender rotational speed.

2.3   MODELING FINISHING MACHINES

2.3.1   SUPERCALENDER

After PM, paper rolls come to finishing machines, on which, during further process-ing, the paper acquires its final properties and shape. These machines include super-calenders (SC) and slitters (Sl). The layout of these devices has much in common: they have an unwinder and a winder, between which the main processing device is located. These installations have the common property that they operate discretely,

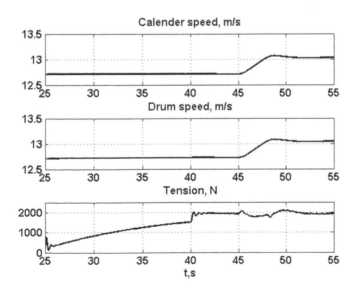

FIGURE 2.50   Tension deviation by PM speed increase with moderate acceleration.

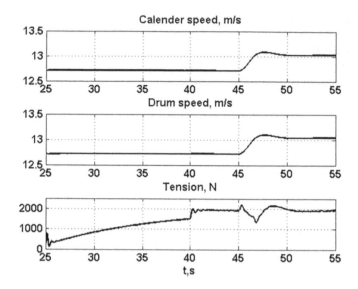

FIGURE 2.51   Tension deviation by PM speed increase with increased acceleration.

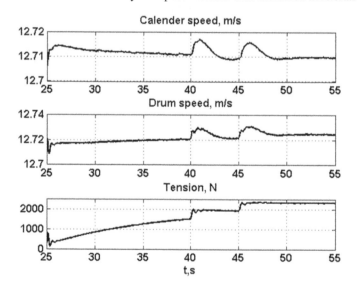

FIGURE 2.52   Tension response for the reference increase.

with stops, but they serve the PM, which operates continuously. To cope with the service, the speed of paper web movement in such machines must be much higher than the speed of the serviced PM. In this section, SC modeling is considered.

   SC is intended to increase the gloss, smoothness, and compaction of paper and equalize its thickness, mainly along the width of the web. The SC consists of 8–12 rolls located one above the other, which the processed paper web sequentially goes around, Figure 2.53. Part of these rolls is metallic, and the rest are stuffed (filled with paper), and the rolls alternate with each other. At the same time, the outer rolls,

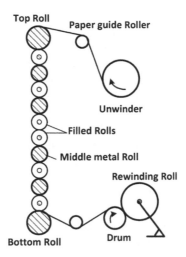

FIGURE 2.53   SC arrangement.

which must withstand high pressure, are metallic. Since the surface of the paper adjoining to the metal rolls acquires a higher smoothness and gloss than the surface adjoining to the stuffed rolls, an even number of rolls is installed on the SC to obtain paper with similar smoothness on both surfaces, while two adjacent stuffed rolls are located in the middle. The paper web, successively passing between the rolls located before two adjacent stuffed ones, all the time comes into contact with the same surface with the metallic rolls. With further passage after two adjacent stuffed rolls, the paper web is already in contact with the metallic rolls with its other surface. With an even number of rolls, the unwinder and the winder are located on one side of the rolls, since the paper on the SC is fed to the upper roll, and exits between the first and second rolls from the bottom.

With a paper web width of 4.2 m, the diameters of the lower and upper rolls can be 800 and 650 mm, respectively, and the intermediate 350–400 mm, the diameters of the stuffed rolls are up to 550 mm, and the thickness of the filling is approximately half the radius. The lower roll is driven in the SC, and the rest rotate, being pressed against each other both by their own weight and by the additional force, so that the maximum pressure between the two lower rolls can reach 350–400 kN/m. The SC speed reaches 1100 m/min. The acceleration values in the SC are also usually larger than those used in the PM. However, as noted in [20], the maximum velocities and pressures cannot be achieved simultaneously due to the relatively fragile design of the stuffed rolls.

Winder (reel) in SC is of the same type as in PM. The unwinder control system has to provide a roll slow rotation in the winding direction for threading of the new web, and also the tension fabrication by all rotational speeds including zero one. Earlier, the DC generators were used for these aims, with the voltage boosters, which supplied the resistors, but in new SC, either DC motors with the reversible thyristor converters or IM with the frequency converters are employed; therefore, the control systems of the SC unwinders are practically the same as those of the uncoilers in cold rolling mills discussed in the first chapter.

The cycle of operation of the SC includes threading the front end of a new roll, setting the rest tension for the unwinder and reel, speeding up to the threading speed and then to the operating speed, decelerating to a stop at the end of rewinding. An emergency stop mode is also provided in case of a web break.

The power of the SC motor depends on the applied pressure, and it is determined by the properties of the processed paper. Earlier in the book, light grades of paper with a width of 4.8 m were considered, so we will take the maximum pressure between the two lower rolls to be $300 \times 4.8 = 1440$ kN. It is noted in [18] that 70%–80% of the power is spent on overcoming the rolling friction between the rolls. The formula for calculating the rolling friction force (2.22) has already been given above. We assume that the compression force changes linearly from 480 to 1440 kN as it moves down, so that when calculating F_{fr}, we can take as the average value of $F_p = 960$ kN for all pairs. Taking the diameters of the metallic rolls (upper, lower, and middle, respectively) as 0.65, 0.8, and 0.4 m, the stuffed rolls 0.55 m, and the value $f_{fr} = 4 \times 10^{-4}$ m [18], we find, with reference to the design in Figure 2.53,

$$F_{fr} = 8 \times 10^{-4} \times 960 \times \left(\frac{1}{0.8} + \frac{1}{0.6} + \frac{8}{0.4} + \frac{12}{0.55} \right) = 34.36 \text{ kN}$$

that, with the speed of calendaring of 750 m/min, gives the power of 430 kW. Taking into account the other losses, for example, in the bearing, the motor demanded power should be not less as $430/0.7 = 614$ kW. The rotational speed of the load shaft is $750/60/0.4 \times 30/\pi = 300$ rpm. Therefore, the torque applied about an SC shaft must be $30 \times 614/\pi/300 = 19.6$ kN-m. In addition, the chosen motor must provide acceleration about 0.5 m/s² and by 1.5–2 times more intense deceleration.

Let's take the IM motor 675 kW, 690 V, 50 Hz, $Z_p = 3$, 6477 N-m, 995 rpm; motor parameters: the rated current 700 A, the non-load current 265 A, the maximal torque M_m, the starting torque and the stating current are equal to 2.1, 0.7 and 6.1 pu, respectively, the moment of inertia 30.2 kg-m², the efficiency 0.965, cos $\varphi = 0.84$. кпд = 96.5 [21]. The motor rotates SC with the help of a gearbox with $i = 3$. The SC moment of inertia can be estimated as 4100 kg-m², therefore, the dynamic torque during speeding up is

$$M_{adin} = \left(\frac{4100}{3^2} + 30.2 \right) \times \frac{12.5 \times 3}{0.4 \times 25} = 1822 \text{ N-m.}$$

It is interesting to compare the power of the selected motor with the real installations. This is not easy, as SCs can vary significantly depending on what grades of paper they are designed for. In particular, in [18] for the similar SC with ten rolls with a width of 4.2 m and a maximum pressure of 250 kN/m, a power of 600 kW is indicated, which approximately corresponds to the selected motor.

The stiffness coefficient is calculated by (1.15), (1.16) with $T_{cr} = 2 \times 2.1 \times 6477 \times 3 = 81.6$ kN-m, $\tau = 80$ N/mm², $l = 3$ m:

$$D = \sqrt[3]{\frac{16 \times 81.6}{\pi \times 80 \times 10^3}} = 0.17 \text{ m}, C = \frac{\pi \times 80 \times 0.17^4 \times 10^9}{32 \times 3} = 2.2 \times 10^6 \text{ N-m/rad,}$$

that is, $\omega_r = \sqrt{\dfrac{2.2 \times 10^6 \times (4100 + 30.2 \times 9)}{4100 \times 30.2 \times 9}} = 92.9$ rad/s, $\omega_{ar} = \sqrt{\dfrac{2.2 \times 10^6}{4100}} = 23.2$

rad/s, $B = \dfrac{2 \times 2.2 \times 10^6 \times 0.07}{92.9} = 3315$ N-m-s/rad.

The **Paper7** model models the main drive of the SC under the assumption that the tensions are kept in all modes. The direct vector control system from the **Paper2a** model is used to control the motor. Since it is necessary to ensure fast deceleration, an active rectifier is used in the DC link, as in the **DTC_shear1** model, but with a two-level converter.

The responses of the drive to a 5% speed reference step without and with a notch filter are shown in Figure 2.54. To implement this mode, the acceleration and deceleration values were increased by 100 times. It can be seen that the presence of a filter significantly reduces the tendency of the system to oscillate. A cycle of SC operation is shown in Figures 2.55 and 2.56. It is assumed that by the beginning of the simulation, the paper web is threaded, and the tensions are set: in the unwinder 1 kN, in the reel 2 kN, the compression force is applied, so that the load torque applied to the motor shaft is $(19,600 - (1000 \times 0.4))/3 = 6400$ N-m. The motor accelerates to the threading speed of 18.75 m/min. At $t = 5$ s, acceleration to the operating speed of 753.6 m/min begins which corresponds to the motor speed of 900 rpm, with an acceleration of 40.5 m/min/s. Operating speed is reached at $t = 27$ s. At $t = 35$ s, the load torque increases by 10% in 3 s, and at $t = 50$ s, the stop process begins with a deceleration rate twice as high as during acceleration. It can be seen from Figure 2.55 that the motor torque during acceleration does not exceed the permissible value; the decrease in the speed of rotation of the SC with increasing load is small and quickly returns to the set value. The plots of the motor and network currents and changes in active and reactive powers are shown in Figure 2.56.

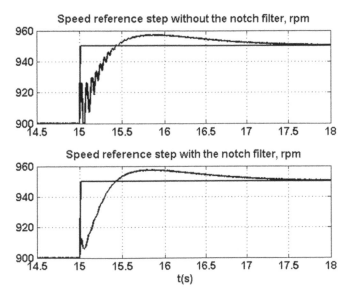

FIGURE 2.54 SC drive response to stepwise speed reference change.

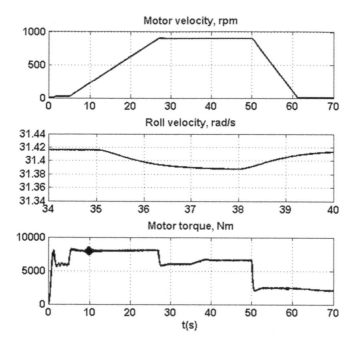

FIGURE 2.55 SC operation cycle.

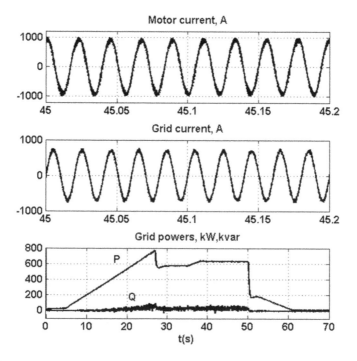

FIGURE 2.56 Currents and powers by SC operating cycle.

It follows from the previous section that the paper rolls with a diameter of up to 2540 mm which are reeled up to the spools with a diameter of 420 mm come to the unwinder. Therefore, the shaft rotational speed should change from $12.5/1.27 \times 30/\pi = 94$ rpm to $12.5/0.21 \times 30/\pi = 569$ rpm. IM are used for all devices in the SC under consideration. For the unwinder, IM from the library *Simscape/Electrical/ Specialized Power Systems* (previous name *SimPowerSystems*) is selected having the data: 160 kVA, 400 V, 50 Hz, 1028 N-m, 1487 rpm ($Z_p = 2$). The motor drives the unwinder with the help of the gearbox with $i = 3.65$. The motor maximal rotational speed is $569 \times 3.65 = 2077$ rpm, that is, it must operate with the flux reduction.

Let's check the dynamic ability of the chosen motor. The roll maximal diameter at the winder is taken equal to 1500 mm, the drum diameter is 600 mm and the reel spool diameter is 100 mm. So, three shipping rolls can be obtained from one parent roll. If to designate D_{1-} the diameter of the roll on the unwinder, at which the winding of the next roll on the reel begins, D_{2-} the diameter of the roll on the unwinder, at which the winding of this roll on the reel ends, then must be $D_1^2 - D_2^2 = 1.5^2 - 0.1^2$. For the first roll, $D_1 = 2.54$ m, hence $D_2 = 2.052$ m; for the second roll, $D_1 = 2.052$ m, hence $D_2 = 1.4$ m, that is, for the third roll $D_1 = 1.4$ m and the shipping roll will have a slightly smaller diameter. The dynamic torque required to accelerate the unwinder with acceleration a (m/s²), on the basis of (2.24), is equal to

$$M_{din} = \left(8862 \times \left(R^4 - 0.21^4\right) + 100\right) \times a/R \qquad (2.28)$$

where R is the value of the roll radius on the unwinder during acceleration, the change of which during acceleration can be neglected. Here, 100 kg-m² is the total moment of inertia of the spool and the motor, reduced to the unwinder shaft. The dynamic torque on the unwinder motor shaft and the rotational speed of the latter when the acceleration value is 0.4 m/s² are shown in Figure 2.57. It can be seen that during the

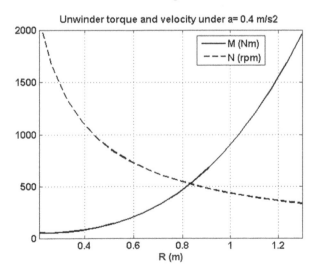

FIGURE 2.57 Dependencies of the dynamic torque and motor rotational speed on the roll diameter for the unwinder.

processing of the first roll, the torque significantly exceeds the allowable one. If we assume that the dynamic torque during acceleration should be approximately equal to the nominal one, assuming that the overload ability is used to create tension and compensate for losses, then during the processing of the first roll, the acceleration should not exceed $0.2\,\text{m/s}^2$, that is, the speeding up time should be $12.5/0.2 \times 63\,\text{s}$, and for the second and third rolls in the cycle, this time can be reduced to $30\,\text{s}$.

To control the unwinder motor, the same direct vector control system is used as for the SC drive, with the difference that the stator flux linkage reference is reduced in relation to N_{nom}/N when the rotational speed N exceeds the nominal value.

The unwinder electric drive is modeled in a simplified way in the **Paper7a** model. The simplification is in the fact that the SC drive is not modeled, but it is assumed that the linear speed of the web at its input is constant. In addition, the inverter's active rectifier is not modeled. It should be noted that the motor parameters used in the control system are set in the **IM_drive, 400 V, 50 Hz** subsystem dialog box and separately in the IM model, which makes it possible, if necessary, to investigate the effect of inaccurate knowledge of the IM parameters on the control quality.

Figure 2.58 shows the changes in the motor rotational speed and torque when the unwinder is turned on at the threading speed, which is assumed to be $1\,\text{m/s}$. The acceleration value is assumed to be 150 rpm/s. It can be seen that the process proceeds quite smoothly.

The unwinding process for an initial radius of $1.026\,\text{m}$ is shown in Figure 2.59. To speed up the simulation at a steady speed, the paper thickness increases by a factor of 30 when this speed is reached until the start of speeding down begins, and this increase occurs smoothly, since otherwise noticeable tension deviations are observed at the moments of thickness change, which, of course, is not the case in fact. It can be seen that the tension is kept with a sufficient degree of accuracy. At the beginning of

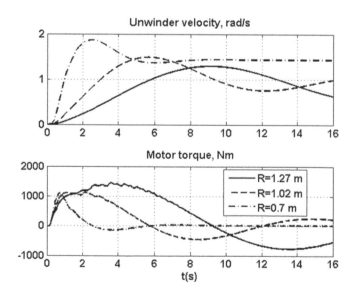

FIGURE 2.58 Unwinder speeding up to the threading speed.

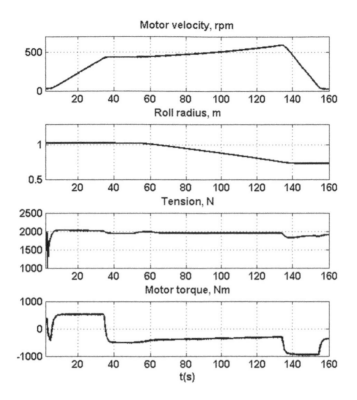

FIGURE 2.59 Unwinding with the initial roll radius 1.026 m.

the start, there are large fluctuations in tension, since the transition from rest tension to tension generation at low speeds is not modeled accurately. To eliminate oscillations, the **Tension_Torque** subsystem is used, with the help of which, at the low speed, the transfer function of the system: tension change ΔT—load change ΔM_l changes, with a smooth transition to natural $\Delta M_l/\Delta T = R$. The load torque component $106\,R^2$ is an estimation of the friction torque in the uncoiler necks.

The IM with a power of 110 kVA at a voltage of 400 V and a rotation speed of 1487 rpm is taken for the winder (reel), which drives the reel drum with help of the gearbox with $i = 3.2$, since the rotational speed of the drum is $12.5/0.3 \times 30/\pi = 398$ rpm. The motor moment of inertia $J_m = 2.3$ kg-m². Coefficients C and B are the same as in PM. The moment of inertia of the drum is 260 kg-m².

The **Paper7b** model also uses the **Tension_Torque** subsystem and increases the paper thickness at a steady speed to reduce the simulation time of the roll winding. The **Control** subsystem provides two options for tension control: indirect, by setting the motor torque, as in the unwinder electric drive, and direct, with measuring the tension of the paper web. In the second variant, the calculated value of the dynamic torque is added to the output of the tension controller when the web speed changes in order to reduce the dynamic error. The loss calculation subsystem **Add_Load** is the same as in the unwinder model, but with numerically reduced losses, since the maximum roll diameter is smaller. Figure 2.60 shows the roll winding cycle when the paper

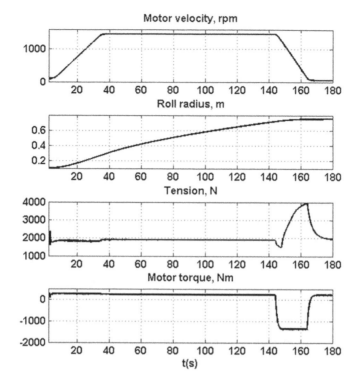

FIGURE 2.60 Winding with the indirect tension controller.

thickness is increased by 20 times. It can be seen that the specified tension is kept with sufficient accuracy, except for the deceleration period at the end of winding, which is caused by the assumed high deceleration rate (36 m/min/s), which the selected motor cannot cope with. It can be shown that when the deceleration rate decreases by 1.5 times, there is no tension surge. Figure 2.61 shows the same process when the tension direct controller is employed (**Manual Switch** tumbler in the **Control** subsystem is in the up position). The processes are almost identical, but it must be borne in mind that the high accuracy of tension control with the indirect method is provided by the assumption that the loss components can be calculated accurately.

In the **Paper7c1** model, the SC electric drives are completely modeled. The active rectifiers for powering the inverters are modeled, with both unwinder and winder inverters being supplied from the same active rectifier. Since the simulation is slow, only fragmentary results are given below. Figure 2.62 shows the process of acceleration from the threading speed to the working speed with indirect tension control. The initial roll diameter is 2.052 m and acceleration is 0.4 m/s² with a paper speed of 12.5 m/s. The plots of powers and currents of the network and changes in the supply voltages of inverters (1200 V for the SC drive and 700 V for the winder and unwinder) during acceleration are shown in Figure 2.63. The THD in the network currents is about 6%, the specified DC voltages are maintained with sufficient accuracy and the reactive power of the network is close to zero.

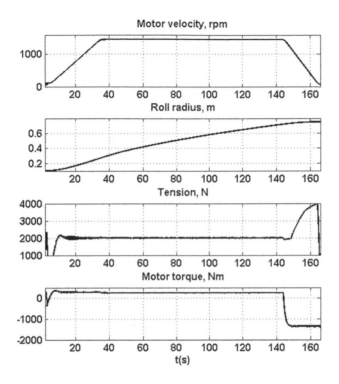

FIGURE 2.61 Winding with the direct tension controller.

The **Paper7c3** model is the same as the **Paper7c1** model but is designed to simulate the end of the roll winding. Measures have been taken in the model to quickly reach the value of the winding roll diameter close to the final one (1.5 m): the acceleration rate is increased to 1.6 m/s² and the paper thickness is increased by 100 times. Before deceleration starts, the paper thickness value returns to the actual value. The deceleration rate is 0.4 m/s², and the tensions on the unwinder and winder are 2000 and 3000 N, respectively. The work with this model demonstrates the difficulty of obtaining accurate values of given tensions with indirect tension control, especially at the instants of change of the SC speed with values of the roll diameter close to the final ones. So, for example, the torque of tension on the winder in this case is 0.9 kN-m, and the total moment of inertia with a full roll is ~2700 kg-m², which, with decelerating of 0.4/0.3 = 1.33 rad/s², gives a dynamic moment of ~3.6 kN-m, so that relatively small errors in the calculation of the dynamic torque lead to much larger, in percentage terms, errors in the implementation of the wanted tension. It also turned out to be necessary during simulation, in order to equalize the dynamics of the drives at the start of deceleration, to introduce first-order inertial elements into the circuits for calculating the derivatives of the given speed, which are placed in the circuits for calculating the dynamic torques of the winder and unwinder and in the circuit for setting of the SC speed, the time constants of which are selected during modeling. Figure 2.64 shows the deceleration process when these constants for the SC and the winder are taken equal to 2 s. The deceleration process starts when the roll diameter on the winder reaches 1.46 m.

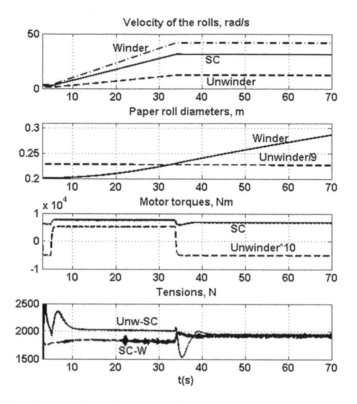

FIGURE 2.62 Process of speeding up from the threading speed to the working speed with indirect tension control.

Tension deviations are reduced by using a direct tension controller for the winder, Figure 2.65, they amount to 2.8–3.2 kN at the beginning of deceleration, while with indirect control they are in the range of 2.5–3.4 kN; there are also no errors in the tension value before the start of speeding down and during its process (equal to −100 and 200 N, respectively, with indirect control, Figure 2.64).

2.3.2 Slitter

Slitter is designed to cut the paper roll produced at PM into rolls of smaller width and often smaller diameter, corresponding to the requirements of the end user. The most common slitter layout is shown in Figure 2.66. The roll of paper produced by the PM is installed on the unwinder. In the process of cutting and rewinding, the paper web passes through a guide roller, a knife block, goes around the first reel drum, and, already cut, is wound into finished rolls. A double-drum reel is usually used in the slitters. The second, along the paper, drum serves to support the roll and to obtain a dense winding of paper into a roll. As the diameter of the roll increases, the pressure of its mass is sufficient to obtain a tight winding, and the pressure from the side of the rider roller decreases. The powers of the motors of the rider and paper-guiding rollers, as well as the knives, are low (e.g., at a speed of 2400 m/min, the sum of the powers

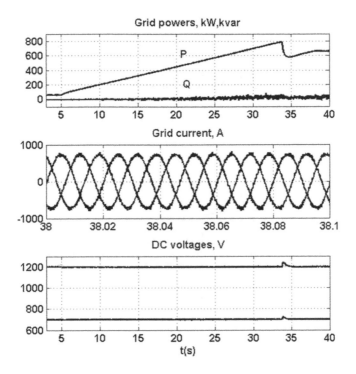

FIGURE 2.63 Powers and currents of the network and changes in the supply voltages of inverters during acceleration.

of both motors of the rider roller is 20 kW, the paper-guiding roller is 40 kW, and the knives are less than 10 kW [22]). These motors must provide a linear speed equal to or slightly higher than the speed of the slitter. They are not considered further.

Assume that paper-grade Newsprint with a density of 49 g/m², thickness (caliper) is 70 mkm, having a mass of 700 kg/m³ is cut. Input (parent) roll width is 6.75 m, maximal diameter is 2.4 m with the spool diameter of 0.6 m. The maximal diameter of the output (shipping) rolls 1.5 m, the slitter speed is 2400 m/min = 40 m/s, the tension maximal value is 2.5 kN, the speeding up and speeding downtimes are 2 min each one, that is, the acceleration/deceleration value is 0.33 m/s². Then, the demanded dynamic torque of the unwinder with the full diameter of 2.4 m can be found from (2.24), (2.28) as $M_{din} = \left(0.5 \times \pi \times 6.75 \times 700 \times \left(1.2^4 - 0.3^4\right) + 200\right) \times \dfrac{0.33}{1.2} = 4270$

N-m. Here 200 kg-m² is the moment of inertia of the motor and spool. It is assumed that the wanted tension value can change by 2.5 times, that is, the tension torque with the full diameter can change from $1 \times 1.2 = 1.2$ kN-m to $2.5 \times 1.2 = 3$ kN-m. Since under speeding up, the dynamic and tension torques have opposite signs, during acceleration with the full roll (if to neglect the additional torques of the small values) the maximal torque $M_{acc} = 4270 - 1200 = 3070$ N-m.

During speeding down, both torques have the same signs; at that, it is necessary to keep in mind that, when speeding down, even the first roll being cut, the mass of the roll being unwound has already decreased. The roll diameter D_{r2} at the end of cutting

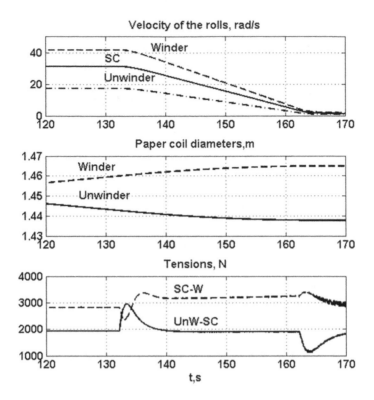

FIGURE 2.64 Process of speeding down in SC with indirect tension control.

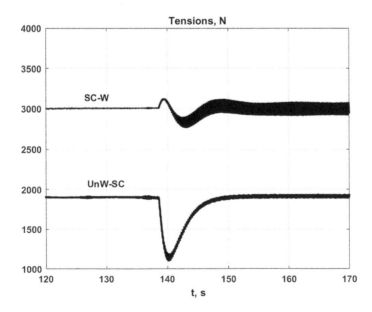

FIGURE 2.65 Tension deviation during speeding down in SC with direct tension control.

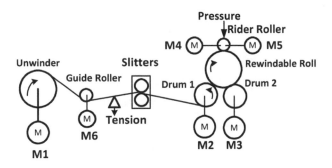

FIGURE 2.66 Slitter layout.

of the first roll can be found from the relation $2.4^2 - D_{r2}^2 = 1.5^2 - 0.1^2$, where 0.1 m is the spool diameter of the winding (being cut) roll, from which $D_{r2} = 1.88$ m, which corresponds to the dynamic torque equal to 2083 N-m. However, since there are very likely cases when the first roll to be cut must have a smaller diameter or it is necessary to brake ahead of time due to an unforeseen situation, we select the motor based on the maximum possible braking torque of 4270 N-m + 3000 N-m = 7270 N-m.

For the IM with $Z_p = 4$ which allows an overload of 1.5 within 2 minutes, the demanded motor power, without gearbox, may be estimated as $(7270/1.5) \times 750 \times \pi/30 = 380$ kW.

Finally the IM is chosen with parameters [21] 355 kW, 690 V, 50 Hz, $Z_p = 4$, 4574 N-m, 741 rpm, the rated current 395 A, the non-load current 192 A, the maximal torque M_m, the starting torque and the stating current are equal to 2.5, 1.2 and 5.9 pu, respectively, the moment of inertia 17.1 kg-m², efficiency 0.953, cos $\varphi = 0.79$.

During the unwinding process, at the roll diameter of 1.02 m, the motor reaches its rated speed, and a flux reduction is required to further increase the speed. The value of the roll diameter D_d, at which the slitter speed reduction should start so that the end of the roll is released at the threading speed, is found from the relation

$$\frac{40^2}{0.66} = \frac{\pi}{4 \times 70 \times 10^{-6}} (D_d^2 - 0.6^2)$$

whence $D_d = 0.76$ m, and deceleration starts at a rotation speed of 1006 rpm.

The **Paper8a** model models the unwinder in a simplified way, assuming that the linear speed on the reel is constant and determined by its setting; the reel lag is taken into account by a first-order element with a time constant of 1 s. In fact, this is the **Paper7a** model, in which the parameters have been changed accordingly. To speed up the simulation, after reaching the set speed (at $t = 150$ s), the paper thickness increases by a factor of 20 and returns to the true value shortly before the start of deceleration.

A cycle of operation with the employment of the direct tension controller and when the reference tension is 2500 N is shown in Figure 2.67. The circuits for the dynamic torque compensation remain in operation and facilitate an operation of the tension controller. It can be seen that the specified tension is kept with sufficient accuracy. Tension surges occur at moments when the paper thickness is artificially

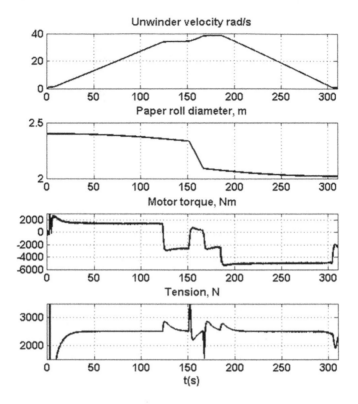

FIGURE 2.67   Cycle of operation of the slitter unwinder.

changed by a factor of 20, which does not actually occur and can be ignored. Small changes in tension also appear at the beginning of speed changes. Let's consider some characteristic points.

At $t = [120...150]$ s, the steady state takes place, with the roll radius $R = 1.17$ m. The motor torque is −2600 N-m. In this case, the tension torque is −2925 N-m, the loss torque is $106 \times 1.17^2 = 145$ N-m, the ventilation loss torque is $0.5 \times 40/1.17 \times 17$ N-m. If to look at the large-scale plot of the speed change, it can be seen that the speed increases with acceleration of 11×10^{-3} rad/s in this range; at this, the moment of inertia is equal to 14,000 kg-m² on average, then the dynamic torque is 154 N-m. Summing up all the above components, we obtain an estimate of the motor torque equal to $-2925 + 145 + 17 + 154 = -2609$ N-m which coincides with the actual value obtained.

In the range of 150–167 s, there is a rapid increase in rotation speed at a constant linear speed, since the thickness of the paper is increased by 20 times. In this case, the acceleration is 0.27 rad/s, and the moment of inertia varies from 13.8×10^3 to 8.8×10^3 kg-m² with an average value of 11.3×10^3 kg-m², so that the average value of the dynamic torque is 3051 N-m. Since in this range the roll radius varies from 1.17 to 1.044 m with an average value of 1.11 m, the average value of the tension torque is −2767 N-m, so that, taking into account the above loss components, the torque should be close to 500 N-m, which can be seen in Figure 2.67.

Above, the value of the dynamic torque during acceleration was found to be 4270 N-m. Since the value of the tension torque is taken equal to $2500 \times 1.2 = 3000$ N-m under simulation, then, taking into account the losses, the motor torque during acceleration should be close to 1.5 kN-m, which can be seen in Figure 2.67. As it can be seen in the figure, the deceleration torque is -5000 Nm, which corresponds to the previously found value of the dynamic torque of 2083 N-m.

The process of the roll cutting when the initial diameter of the roll in unwinder is 1.8 m is shown in Figure 2.68 (**Paper8a1** model). Since the moment of inertia of this roll is small already, and the moment of inertia of the roll in winder, at winding beginning, is small yet, the acceleration value is doubled, to $0.66\,\text{m/s}^2$. At $t = 100\,\text{s}$, the paper thickness is increased tenfold, and returns to its original value when the roll diameter in the unwinder is 0.94 m. As the roll diameter decreases, the speed of the unwinder motor increases and starts to exceed the nominal value, so that the motor flux reduces. When the diameter reduces to 0.84 m, deceleration begins. The tension is kept equal to the specified one.

Consider the reel drives. In accordance with the existing recommendations, the sum of the powers of both double-drum reel motors is accepted by 30% more than the power of a single-drum reel with the same parameters. In our case, taking into account the increase in the speed of the slitter compared to the speed of the SC and the increase in the width of the paper, the power of each motor should be equal to

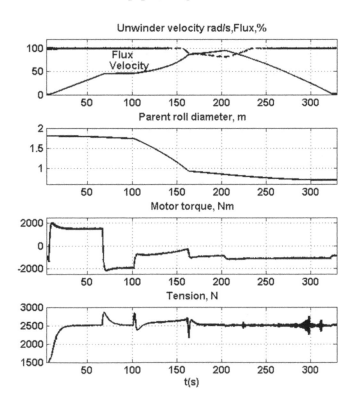

FIGURE 2.68   Roll cutting when the initial diameter of the roll in unwinder is 1.8 m.

$110 \times (2400/750) \times (6.75/4.8) \times 1.3/2 = 320$ kVA. With a drum diameter $D_d = 2R_d = 0.6$ m, the motor speed should be 1274 rpm. The motors are accepted with parameters (taking in mind a unification of the supplying voltages): 280 kW, 690 V, 50 Hz, $Z_p = 2$, 1491 rpm, the rated current 289 A, the non-load current 112 ×, the maximal torque M_m, the starting torque and the stating current are equal to 2.4, 0.7 and 6.2 pu respectively, the moment of inertia 4.6 kg-m², efficiency 0.962, cos $\varphi = 0.84$. The drum moment of inertia is 260 kg-m².

The stiffness coefficient C is calculated by (1.15), (1.16) under $T_{cr} = 2.4 \times 1793 \times 3 = 12.9$ kN-m, $\tau = 80$ N/mm², $l = 3$ m:

$$D = \sqrt[3]{\frac{16 \times 12.9}{\pi \times 80 \times 10^3}} = 0.09 \text{ m}, C = \frac{\pi \times 80 \times 0.09^4 \times 10^9}{32 \times 3} = 0.17 \times 10^6 \text{ N-m/rad},$$

that is, $\omega_r \approx \sqrt{\dfrac{0.17 \times 10^6}{4.6}} = 192$ rad/s, $B = \dfrac{2 \times 0.17 \times 10^6 \times 0.07}{192} = 124$ N-m-s/rad.

The torques of the first (M2—Figure 2.66) and the second (M_3) motors are determined by the following relationship, respectively

$$M_{e1} = T_r R_d + M_{fd1} + F_{dr1} R_d + M_{din}$$

$$M_{e2} = M_{fd2} + F_{dr2} R_d + M_{din} (2.29)$$

where T_r is the tension, M_{fd} is the frictional torque in the drum bearings, M_{din} is the torque demanding for the drum acceleration, F_{dr} is the tangential contact force between the drum and the paper roll. The frictional torque

$$M_{fd} = f_c Q_c d_d / 2 (2.30)$$

where f_c is the friction factor in the bearing journal, $f_c = 0.004$, Q_c is the weight of the drum and of the roll weight share for one drum, d_d is the journal diameter which is taken equal to 0.4 m. The drum mass is taken equal to 2500 kg, the roll mass $m_r = \pi b \gamma_p R_r^2 = 14{,}836 R_r^2$. Then

$$M_{fd1} = M_{fd2} = 0.004 \times \left(2500 + 0.5 \times 14836 R_r^2\right) \times 9.8 \times 0.2 = 19.6 + 58.2 R_r^2 \text{ N-m}.$$

The sum of tangential forces $F_{dr1} + F_{dr2}$ is spent to overcome the rolling friction between the drums and the roll F_{fr} and to compensate for dynamic torques when the slitter linear speed changes and when the kinetic energy of the roll changes during winding.

From (2.22), the force of one drum to overcome the rolling friction is

$$F_{fr} = 2 f_{fr} F_p \left(1/D_d + 1/D_r\right)/2 \tag{2.31}$$

where f_{fr} is the rolling friction coefficient that is equal to 0.0022 m, F_p is the total pressure on the drums from the weight of the roll and the pressure of the rider roller, N. As long as the mass of the roll is small, the required pressure is provided by the rider roller the pressure on which changes with a change in the diameter of the roll. The full value of this pressure is taken equal to 4 kN/m. When the wanted pressure is provided by the weight of the roll, the pressure on the roller is reduced to small values sufficient to ensure that the roller is pressed against the roll to measure its diameter, and the total pressure can be calculated as

$$F_p = 9.8 \times 0.25\pi D_r^2 b\gamma_p \cos\alpha, \qquad (2.32)$$

$$\sin\alpha = \frac{k}{D_r + D_d} \qquad (2.33)$$

where k is the distance between drum centers. If to take $k = 0.65\,$m, one can see that the wanted pressure is reached with the roll diameter of 0.9 m. The change of the force of one drum to overcome rolling friction, depending on the diameter of the roll, is shown in Figure 2.69.

The M_2 motor operates in speed control mode. Its rotational speed is determined by the circumferential speed of the reeled up roll. The rotational speed of the M_3 motor is set slightly higher (by 1%–3%) than the rotational speed of M_2. Since the paper on the roll has some degree of freedom, the higher rotational speed of M_3 provides additional movement of the paper to be wound in the direction of winding and thus contributes to tighter winding. Increasing the speed of M_3 increases its load torque and, accordingly, reduces the load torque of the motor M_2. Distribution of the torques between the motors is determined by the value $K_{sh} = M_3/(M_2 + M_3)$ where M_2 and M_3 are the torques of the motors and are defined by the relative difference $a = \dfrac{M_3 - M_2}{M_2 + M_3}$, so that for a given value of K_{sh} the value $a = 2K_{sh} - 1$.

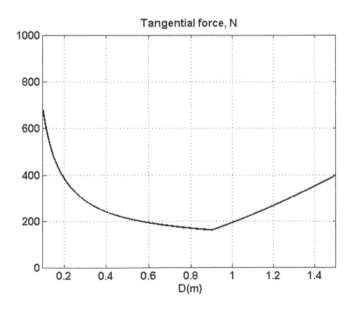

FIGURE 2.69 Dependence of the tangential force on the roll diameter.

The greater the speed difference, the greater the value of K_{sh}. It is assumed that with a difference of 3%, the value $K_{sh} = 1$ is reached.

The organization of the load distribution and motor torque control is shown in Figure 2.70. The output of the M_2 speed controller determines the total required torque of the motors to obtain the desired rotational speed. This torque is distributed to the control systems of M_2 and M_3 motors depending on the desired value of K_{sh} which in turn depends on the current radius of the roll, usually decreasing with increasing radius. These distributed quantities determine the values of the motor current components i_{T2} and i_{T3}, which are proportional to the torques of these motors.

In the **Paper8b** model, the simulation of the process in the winding paper roll is performed in the **Shipping_roll** subsystem. It calculates the roll radius and the moment of inertia, as in the previous models, and the moment of inertia of a spool with a diameter of 0.1 m is taken equal to $J_0 = 20$ kg-m². To control the speed of rotation of the roll and the dynamic torque required for this, a slightly modified scheme is used, shown earlier in Figure 2.27, since now the dynamic torque is defined as $M_{dinr} = \dfrac{d}{dt}(J_r \omega_r)$, Figure 2.71. The formulas for calculating the dynamic torque, taking into account the change of the moment of inertia and the speed of rotation of the roll at a constant linear speed V, were given above (2.23–2.27).

Figure 2.72 shows the response of the model to changes in the linear speed and radius of the roll and the results of calculating the dynamic torque using the

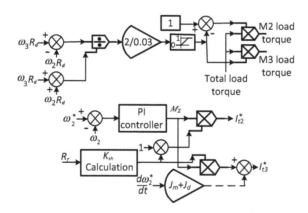

FIGURE 2.70 Block diagram of the load distribution and motor torque control in the two-drum reel.

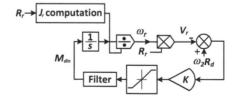

FIGURE 2.71 Block diagram of the circuits for the speed rotation control of the roll and the dynamic torque.

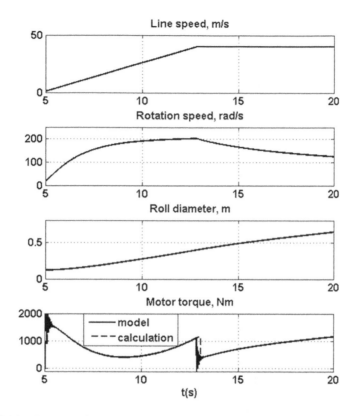

FIGURE 2.72   Processes in the double-drum reel.

above formulas. It can be seen that the results practically coincide, which confirms the possibility of using the developed model to simulate the winding process in the double-drum reel.

Figure 2.73 shows the process that confirms the possibility of such control of the load distribution between the motors. The motors accelerate with a high acceleration corresponding to a linear acceleration of 1 m/s. The coefficient K_{sh}, initially equal to 0.7, at $t = 70$ s decreases to 0.35 (accordingly, a from 0.4 to -0.3), that is, M_2 becomes more loaded motor. Table 2.1 shows the values of the torques of the motors at times of 40 s (acceleration), 70 s, and 100 s. It can be seen that the specified values of K_{sh} are achieved.

One cycle of roll winding is shown in Figure 2.74. The motors speed up with an acceleration of 1 m/s². The coefficient $K_{sh} = 0.7$, so the torques $M_3 > M_2$. After reaching the specified speed (at $t = 50$ s), the paper thickness is increased by a factor of 5, which accordingly reduces the time of operation at a steady speed during simulation. As the radius increases, the coefficient K_{sh} decreases, and the M_2 torque begins to exceed the M_3 torque. When the roll radius R_r reaches 0.67 m (at $t = 147$ s), the thickness value returns to the true value, and at $R_r = 0.69$ m deceleration begins. At the same time, it turned out that due to the small value of K_{sh}, the M_3 motor control system could not create the torque required to slow down the motor, so it was necessary

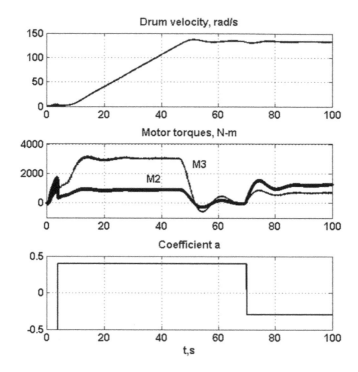

FIGURE 2.73   Control of the load distribution.

TABLE 2.1
To Illustrate the Validity of the Control System

Time (s)	40	70	100
M_2 torgue (N-m)	898	495	1274
M_3 torgue (N-m)	2159	1216	720
K_{sh}	0.71	0.71	0.36
A	0.42	0.42	−0.28

to introduce an additional signal that is proportional to the dynamic torque of the M_3 motor with the drum (moment of inertia is 260 kg-m²) which increases the value of the torque component I_{f3} of M_3 current.

The **Paper8c1** model contains both the unwinder model and the two-drum model, that is, they can be considered as complete models of the slitter electric drives, if to neglect the electric drives of the guide and rider rollers and knives, which have a small power. The unwinder and winder models are **Paper8a** and **Paper8b** models, respectively. Since simulation in these models is slow, only fragments of the processes are presented below. The plots of the first 150 s of the rewinding process are given in Figure 2.75 when the initial diameter of the parent roll is 2 m. The motors speed up to the linear speed of the paper web of 40 m/s with an acceleration of 0.33 m/s² (at the very beginning of acceleration, the acceleration is 0.5 m/s²).

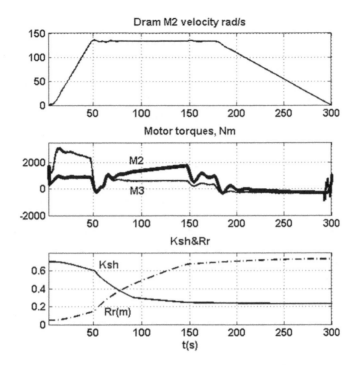

FIGURE 2.74 One cycle of roll winding.

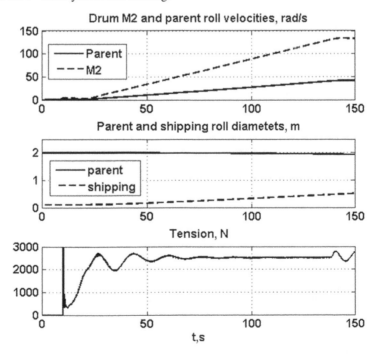

FIGURE 2.75 Start of the winding process.

Acceleration ends at $t = 140$ s. The end of acceleration is accompanied by some short-term increase in tension which is caused by different dynamics of the electric drives. A virtual change of the thickness of the paper web was not performed. At $t = 150$ s, the diameters of the unwinding and winding rolls are 1.936 and 0.513 m, respectively. At this, the condition of conservation of the paper volume holds: $2^2 - 1.936^2 = 0.513^2 - 0.1^2$. Since the diameter of the roll on the reel is less than the value taken for the start of deceleration (1.38 m), it does not occur. The torque of the unwinder motor is approximately 100 N-m, since the negative tension torque is compensated by the positive dynamic torque. The torques of the reel motors M_2, M_3 at $t = 140$ s are 700 and 1100 N-m, respectively.

An active rectifier generating a 1200 V DC voltage is fed from the 6.3 kV mains through a 6300/750 V transformer and a reactor with an inductance of 0.2 mH. Figure 2.76 shows plots of changes in the active and reactive powers of the network, as well as the network current for the same conditions, as in Figure 2.75. It can be seen that the power consumed from the network reaches 260 kW to the end of the acceleration: the reactive power is low. THD in the mains current is approximately 6.5%, while THD in the mains voltage on the primary winding of the transformer under the selected short circuit power of the network reaches 4.5%. It can be seen from the **DC** oscilloscope that the specified DC voltage is kept with a high degree of accuracy.

The end of rewinding of a roll is shown in Figure 2.77. A number of changes were made in the model to reach the end of the rewind quickly, in particular, the acceleration was increased and the virtual paper thickness was increased over a significant length of the roll, the **Paper8c2** model. At $t = 150$ s, a steady state is reached with the

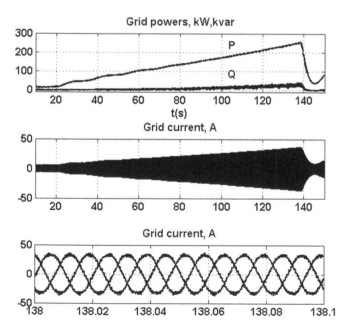

FIGURE 2.76 Active and reactive powers of the network and the network current during start of the winding process.

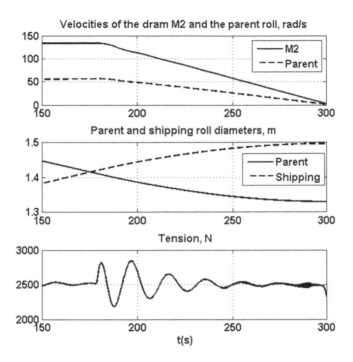

FIGURE 2.77   End of rewinding of the shipping roll.

actual paper thickness and the reel diameter of 1.38 m. At $t = 175$ s, the reel diameter reaches 1.42 m, at which the speeding down starts. It can be seen that at the beginning of deceleration, there is a noticeable change in tension of the order of 13%. At $t = 300$ s, the motors stop with the wanted diameter of the shipping roll of 1.5 m.

REFERENCES

1. Power Requirement of Fourdrinier Machines. *TAPPI*, Vol. 45, No. 2, February 1962.
2. EATON. *Paper Machines Drives*. Eaton Corporation, United States, 1997.
3. Valenzuela, A., Bentley, J. M., Lorenz, R. D. Evaluation of Torsional Oscillations in Paper Machine Sections. *IEEE Transactions on Industry Applications*, Vol. 41, No. 2, March/April 2005.
4. Valenzuela, M. A., Bentley, J. M., Villablanca, A., Lorenz, R. D. Dynamic Compensation of Torsional Oscillation in Paper Machine Sections. *IEEE Transactions on Industry Applications*, Vol. 41, No. 6, November/December 2005.
5. Valenzuela, M. A., Bentley, J. M., Lorenz, R. D. Computer-Aided Controller Setting Procedure for Paper Machine Drive Systems. *IEEE Transactions on Industry Applications*, Vol. 45, No. 2, Marth/April 2009.
6. Bortsov, Y. A., Sokolovsky, G. G. *Thyristor Electric Drive Systems with Elastic Connections* (in Russian). Energy, Leningrad, 1979.
7. Haikola.M., No gear required. *ABB Review*, N4, pp. 12-15, 2009.
8. Lopera, J. M., Granda, A., Linera, F. F., Vecino, G., Díaz, A. Practical Speed and Elongation Measurement, Using Encoders, for a Temper Mill. 2012 IEEE Industry Applications Society Annual Meeting. Las Vegas, NV.

9. Valenzuela, M. A., Bentley, J. M., Lorenz, R. D. Sensorless Tension Control in Paper Machines. *IEEE Transactions on Industry Applications*, Vol. 39, No. 2, Marth/April 2003.

10. Ramírez, G., Lorenz, R. D., Valenzuela, M. A. Observer-Based Estimation of Modulus of Elasticity for Papermaking Process. *IEEE Transactions on Industry Applications*, Vol. 50, No. 3, May/June 2014.

11. Baryshnikov, V. D., Kulikov, S. N. *Automated Electric Drives of Paper-Making Machines (in Russian)*. L. Energoizdat, 1982.

12. Bose, B. K. *Power Electronics and Variable Frequency Drives*. IEEE Press, New York, 1996.

13. Wedel, G. L., Timm, G. L. *Drive Power and Torque in Paper Machine Dryers*. Kadant Johnson Inc., 2008.

14. Valenzuela, M. A., Bentley, J. M., Lorenz, R. D. Condensate Effects on Power and Torque Requirements during Starting of Dryer Sections. *IEEE Transactions on Industry Applications*, Vol. 47, No. 4, July/August 2011.

15. Ramírez, G., Valenzuela, M. A. Online Estimation of the Condensate Load in Dryer Cylinders during Section Starting. *IEEE Transactions on Industry Applications*, Vol. 48, No. 5, September/October 2012.

16. Honaker, D., Jones, W. V. Paper Machine Dryer Section Tuning. *IEEE Industry Applications Magazine*, Vol. 20, No. 2, March/April, 2014.

17. Valenzuela, M. A., Bentley, J. M., Lorenz, R. D. Dynamic Online Sensing of Sheet Modulus of Elasticity. *IEEE Transactions on Industry Applications*, Vol. 46, No. 1, January/February 2010.

18. Eidlin, I. Y. *Paper-Making and Finishing Machines* (in Russian). Publishing House "Forest industry", M. 1970.

19. Valenzuela, M. A., Carrasco, R., Sbarbaro, D. Robust Sheet Tension Estimation for Paper Winders. *IEEE Transactions on Industry Applications*, Vol. 44, No. 6, November/December 2008.

20. Holik, Herbert (Ed.). *Handbook of Paper and Board*. Wiley-VCH Verlag GmbH & Co, Germany, 2006.

21. ABB. *High Voltage Engineered Induction Motors*. Technical Catalog, 2018.

22. Fransen, P. Upgrade of Paper Machine Winders with AC Drives and the Synchronous Rectifier. Conference Record of 1996 Annual Pulp and Paper Industry Technical Conference, 1996.

3 Simulation of the Material Handling Equipment

3.1 PICK-AND-PLACE MACHINES

3.1.1 TRAVELING CRANES

Traveling cranes are designed for mass-handling operations. They are a span structure of a lattice construction that rests on two supports connected to wheels moving along the rail track. Along the span, a trolley moves along the rail track; on the trolley, a drum of hoisting equipment is installed on which a rope is wound, at the end of which there is a loading device: grab, grip, and so on. Next, the traveling crane with the following data will be considered: span length $L = 100$ m, trolley weight without load 110 t, 4 drive wheels with a diameter of $D_r = 0.9$ m, gauge 4.8 m, trolley length 13.5 m, travel speed 240 m/min, load capacity $Q_l = 300$ kN, lifting height 35 m, lifting/lowering speed $v_l = 72$ m/min, drum diameter 1.1 m, drum length 2 m, moment of inertia of the drum $J_d = 625$ kg-m². Figure 3.1 shows a general view of the considered traveling crane where: L is the span, L_g is the length of the bridge, h is the lifting height of the grab, l is the length of the rope, θ is the angle of deviation of the rope from the vertical when the trolley moves, 4 is the trolley, 5 is the driver's cab and electrical equipment, 6 is the grab or other transportable load, M1, M2 are axles of the trolley and their driven motors, M3 is the drum with a rope and its driven motor, S1, S2 are travels of the trolley and load, respectively, m_l—mass of the load.

The peculiarity of the traveling crane operation is that after its loading or emptying, the load-receiving device, for example, a grab, begins to rise, and at the same time the movement of the trolley begins to save time. Thus, the load is a pendulum of variable length, the suspension point of which moves according to a certain law.

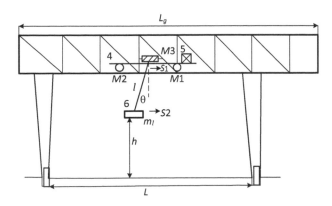

FIGURE 3.1 General view of the traveling crane.

DOI: 10.1201/9781003394419-3

This circumstance leads to the swinging of the load, which is an unpleasant phenomenon for various reasons. Therefore, the design of the electric drive must take this fact into account. As for the movement of the traveling crane itself, this is an adjustment travel and usually takes place without a load. This factor is not considered further.

The motors for the considered traveling crane are chosen in the following. The hoist motor power is

$$P_l = \frac{Q_l v_l}{\eta} = \frac{300 \times 72}{60 \times 0.85} = 424 \text{ kW}.$$

Here $\eta = 0.85$ is the motor efficiency. The drum rotational speed is $[(72/60)/0.55] \times 30/\pi = 21$ rpm. The IM having 4 pole pairs $(Z_p = 4)$ is selected, with parameters: 400 V, 450 kW, 744 rpm, the rated torque 5775 N-m, the maximal torque 2.6 pu, $J_m = 26$ kg-m^2 that rotates the drum with two-stage gearbox, $i = 32.42$. The device total moment of inertia reduced to the motor shaft is calculated by the formula

$$J_l = J_m + J_i + \frac{J_d}{i^2} + \frac{m_l D_d^2}{4i^2} \tag{3.1}$$

where J_i is the gearbox moment of inertia which is taken as $J_i = 0.2 J_m$. We have

$$J_l = 1.2 \times 26 + 625/32.42^2 + 0.25 \times (3^5/9.81) \times 1.1^2/32.42^2 = 40.6 \text{ kg-m}^2.$$

If to take a speeding up time to the maximal speed equal to 1.2 s, the value of the dynamic torque is $M_{dinl} = 40.6 \times \dfrac{1.2 \times 32.42}{0.55 \times 1.2} = 2393$ N-m. Based on the found power, the torque of resistance is $M_{stl} = 424 \times 0.55/(1.2 \times 32.42) = 5.94$ kN-m.

Consider one operation consisting of the stages of loading—lifting the load to a given height—moving the load to the place of unloading—lowering the load. The duration of the operation can be represented as $T_l = t_1 + 2t_2 + 2t_3 + 2t_4 + t_5$ where t_1 is the loading time taken equal to 10 s, $M_{stl} = 0$; t_2 is the time of the acceleration or deceleration with the load $(t_2 = 1.2$ s$)$ when the torques M_{stl} and M_{dinl} are added (acceleration during lifting and deceleration during lowering); t_3 is the same when the toques M_{stl} and M_{dinl} are subtracted; t_4 is the time of load moving vertically, at the steady speed, with the torque M_{stl}, taking the actual hoisting height equal to 32 m,
$t_4 = \dfrac{32 - 2 \times 0.5 \times 1.2 \times 1.2}{1.2} = 25.5$ s; t_5 is the time of the trolley travel when $M_{stl} = 0$.
If to take 0.7 of the span as a minimum possible path then $t_5 = 70/4 = 17.5$ s, $T_l = 83.3$ s, and rms of the motor torque is

$$M_{rms} = \sqrt{\left[2.4 \times \left(8.33^2 + 3.55^2\right) + 2 \times 25.5 \times 5.94^2\right]/83.3} = 4.9 \text{ kN-m}$$

that is less than the motor-rated torque. The maximal motor torque is $2.5 \times 5775 = 14.4$ kN-m which is larger than the torque at the acceleration during load lifting that was computed above and is equal to 8.33 kN-m. The other motor data: efficiency $= 96.2$, $\cos \varphi = 0.83$, $I_{nom} = 744$ A, the starting current 6 pu.

The resistance to the movement of the trolley is determined by the friction forces in the wheel journals, the friction of the wheel rolling along the rail, and the friction of the flanges:

$$F_{l1} = \frac{G_m}{D_r}\left(2\mu + fd_j\right) \times k_p \tag{3.2}$$

where G_m is the trolley weight with the load, N, μ is the coefficient of rolling friction of the wheels on the rail, $\mu = 0.001$ m, f is the coefficient of friction in the journal of a wheel with a diameter d_j, we take $f = 0.02$, $d_j = 0.2$ m, k_p is the coefficient of friction of the flanges, $k_p = 1.45$. Additional movement resistance forces are: F_{l2} is the resistance from the track slope that occurs when the bridge deflects, $F_{l2} = G_m \sin\gamma_{gr}$ where $\gamma_{gr} = 0.002$ is taken, and F_{l3} is the resistance from the wind load. Since the latter value depends on many factors, for its approximate consideration, $F_{l3} = 0.1\ F_{l1}$ is accepted. So $F_l = 1.4 \times 10^6\left(0.002 + 0.02 \times 0.2/0.9\right) \times 1.45 \times 1.1 + 1.4 \times 10^6 \times 0.002 = 17.2$ kN. Taking the gear efficiency equal to 0.85, the motor static power is equal to 17.2 × 240/0.85/60 = 81 kW.

Selected motors must provide the specified value of the trolley acceleration which is taken equal to 0.8 m/s², which corresponds to the dynamic force $(110 + 30) \times 0.8 = 112$ kN. However, the force used to accelerate the trolley must be less than the cohesive force, which for an unloaded trolley, with the wheel-rail traction coefficient of 0.14, is equal to $1100 \times 0.14 = 154$ kN. Taking the safety factor equal to 1.1, the maximum allowable dynamic force is $154/1.1 - 17.2/0.85 = 120$ kN. The required maximum power is $(154/1.1) \times 4 = 560$ kW. Assume that every two wheels are driven by IM with $Z_p = 4$, 200 kW, 400 V, 50 Hz, 742 rpm, the rated torque is 2576 N-m, the maximum torque is 2.6 pu, the starting torque is 1.6 pu, the starting current 7.4 pu, $\cos = 0.79$, efficiency 0.95, the rated current is 385 A, $J_m = 11.3$ kg-m² [1]. IMs rotate the wheels through the gearbox $i = 8.1$. Therefore, under rated conditions, the motors provide the travel speed of $(742/8.1) \times (\pi/30) \times 0.45 = 4.3$ m/s with the rated torque of 5152 N-m. In order to produce the traction force of $154/1.1 = 140$ kN, the total motor torque must be $140 \times 0.45/8.1 = 7.78$ kN-m which is provided under the allowable overload of 1.52.

The equivalent torque of the motors for the most severe scenario is computed when the lifting and lowering of the load is carried out while the trolley is moving. At this, the static torque is $(17.2 \times 0.45)/(0.85 \times 8.1) = 1.12$ kN-m, and the dynamic torque $(112 \times 0.45)/8.1 = 6.22$ kN-m. The duration of the trolley operation cycle can be written as $T_c = t_1 + t_2 + t_3 + t_4$ where t_1 is the time of loading or unloading, taken equal to 10 s, $M_m = 0$; t_2 is the acceleration time ($t_2 = 5$ s) at which the static and dynamic torques are summed up, $M_m = 1.12 + 6.22 = 7.34$ kN-m, and a travel is $0.5 \times 4 \times 5 = 10$ m; t_3 is the deceleration time at which these torques are subtracted (5 s), $M_m = 1.12 - 6.22 = -5.1$ kN-m; t_4 is the time when the trolley moves with the steady speed with the torque of 1.12 kN-m, $t_4 = \dfrac{100 - 2 \times 10}{4} = 20$ s, that is, $T_c = 40$ s. Then

$$M_{mrms} = \sqrt{\left[5 \times 7.34^2 + 5 \times 5.1^2 + 20 \times 1.12^2\right]/40} = 3.26 \text{ kN-m}$$

that is less than the sum of the motor-rated torques.

The **Crane1a** model simulate a hoisting electric drive with an active rectifier. To IM control, the DTC system is used which has been used already in previous models. Since the drive operates in reversing duty with regeneration, an active rectifier is used as a DC source for the inverter, as, for example, in the **DTC_shear1** model. The evaluation of the components of the stator flux linkage vector is performed using the same relationships as, for example, in the **Paper8a** model, equations (1.17)–(1.20). These components are also used for the calculation of the motor torque. However, the simulation showed that there are combinations of IM parameters at which the motor start does not occur from a state of rest when using the torque values calculated in this way. Therefore, the **Torque estimator** subsystem has been added to the model in which the torque is calculated using the following equations [2]:

$$\frac{d\Psi_{qr}}{dt} = -R_r I_{qr} + \omega_r \Psi_{dr} \tag{3.3}$$

$$\frac{d\Psi_{dr}}{dt} = -R_r I_{dr} - \omega_r \Psi_{qr} \tag{3.4}$$

$$T_m = \Psi_{qr} I_{dr} - \Psi_{dr} I_{qr} \tag{3.5}$$

This torque value is only used during the initial start-up of the drive.

To hold the load when the IM is stopped, a mechanical brake is used, which is switched on and off at low rotation speeds. When the brake is turned on, it takes on the weight of the load, and the IM load torque becomes equal to zero. The models measure the lifting height of the load by integrating the circumferential speed of the winch drum.

Figure 3.2 shows the process of lifting and then lowering the load. Lifting starts at $t=0.4$ s, and the lifting speed of the load reaches 1.2 m/s after 1 s. When the lifting height reaches 31.4 m, deceleration starts, and the load stops at the height of 32 m, and the brake is applied. At $t=40$ s, the lowering of the load begins; at $t=66.2$ s, when the load is at the height of 0.6 m, the motor torque increases sharply to slow down its further decrease, and at $t=67.1$ s, the load reaches zero.

The trolley movement mechanism has the peculiarity that the load on the shafts is significantly different, since on one side of the trolley, the driver's cab and electrical equipment are placed. Therefore, torques created by electric motors also are different. In the **Crane1b**, **Crane2b** models, both motors are connected in parallel and are powered by a single inverter. The DTC system is used to control the inverter, but for a change, instead of the previously used system, this model uses the **DTC** control subsystem from the **AC4** library model. The moment of inertia of the motor itself is taken into account in motor models, and the moment of inertia of the trolley is taken into account in the torques of resistance to movement. There are two ways to take into account the trolley torques. In the first model, the load torque of each motor is taken equal to $M_{ci} = k_i \left(J_l \dfrac{d\omega_1}{dt} + M_l \right)$, J_l is the moment of inertia of the trolley with the load reduced to the motor shaft, $J_l = \dfrac{140 \times 10^3 \times 0.45^2}{8.1^2} = 432$ kg-m², M_l is the

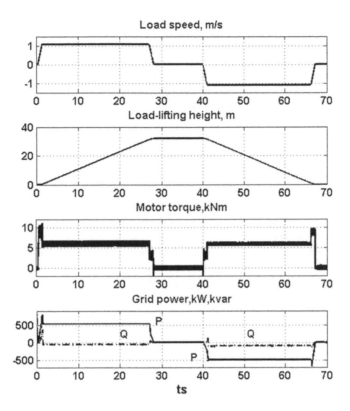

FIGURE 3.2 Load lifting and lowering.

torque of the resistance to trolley movement that was calculated above as 1120 N-m, $i = 1,2$, $k_1 + k_2 = 1$. It is taken $k_1 \geq k_2$, $k_1 = 0.55$, $k_2 = 0.45$, that is, M1 load is 20% larger, and the trolley speed is determined by more loaded shaft. It can be noted that the same simulation results can be obtained by simply setting the corresponding different values of the moments of inertia of the motors and their torques of resistance, but in the accepted model, it is easier to study the processes with changes of the moment of inertia of the trolley and load distribution.

In the second model, the equation of the trolley motion

$$J_l \frac{d\omega_l}{dt} = M_1 + M_2 - M_l \tag{3.6}$$

is solved where M_1 and M_2 are torques developed with the motors and ω_l is the trolley-reduced velocity. Since it must be equal to the average rotational speed of motors, their load torques are given as

$$M_{ci} = k_i K \left(\frac{\omega_1 + \omega_2}{2} - \omega_l \right) \tag{3.7}$$

where K is a sufficiently large number.

Figures 3.3 and 3.4 show the processes of acceleration, operation at a steady speed, and deceleration for both models. It can be seen that they are almost the same. It is seen on the scope **Motor 1** that there is a difference of about 0.2% between the shaft rotational speeds; it means that one of the wheel pairs will spin. The specified ratio of motor torques is kept. So, for example, for the second option during acceleration, the motor torques are 4120 N-m and 3400 N-m, respectively, that is the torque M1 is 21% larger.

The **Crane 1c** model simulates the joint operation of the hoist and trolley electric drives. The inverters are powered by a common active rectifier with an output voltage of 750 V. The graphic display shows the movement of the load online. To reduce the simulation time, it is assumed that the lifting height is 10 m, and the travel of the load is 50 m. The lifting of the load and the movement of the trolley begin at the same time. When the designed travel distance of 37 m is reached, the load-lowering and the slowing-down of the trolley begin. As can be seen from the graph on the display and in Figure 3.5, at the last stage of lowering, the trolley is at rest, having covered a distance of 50.5 m. It is seen from Figure 3.5 that the active power consumed from the grid reaches, for a short time, the value of 1 MW and the reactive power is close to zero.

The considered models did not take into account the vibrations of the suspended load, the presence of which has already been mentioned. If denoted as θ the angle of deviation of the rope from the vertical (Figure 3.1), then the change of this angle when the trolley moves is described by the equation [3]

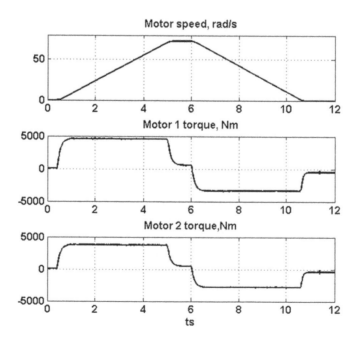

FIGURE 3.3 Processes of acceleration, operation at a steady speed, and deceleration for the **Crane 1b** model.

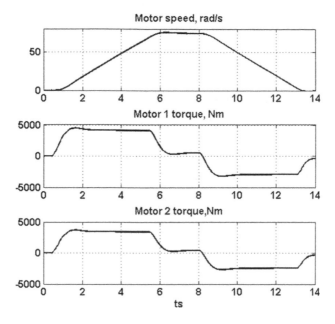

FIGURE 3.4 Processes of acceleration, operation at a steady speed, and deceleration for the **Crane 2b** model.

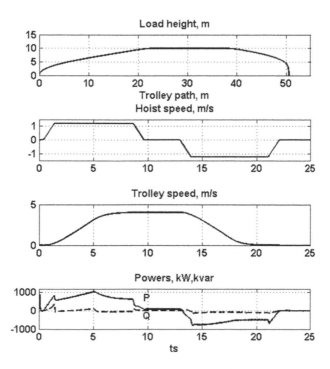

FIGURE 3.5 Joint operation of the hoist and trolley electric drives.

$$\ddot{\theta} + \frac{\ddot{S}}{l}\cos\theta + 2\dot{\theta}\frac{\dot{l}}{l} + \frac{g}{l}\sin\theta = 0. \tag{3.8}$$

Here S is the trolley travel. When deriving (3.8), it is assumed that the rope is weightless and does not stretch, and the load is a material point. When using this equation, it is usually assumed that the angle θ is small, then (3.8) reduces to (3.9)

$$\ddot{\theta} + 2\dot{\theta}\frac{\dot{l}}{l} + \frac{g}{l}\theta = -\frac{\ddot{S}}{l}. \tag{3.9}$$

In (3.9), the second term is often neglected. The deviation of the load from the vertical causes an increase in the resistance to movement of the trolley, the force of this resistance F_r can be estimated as [4]

$$F_r = m_l l \left(\ddot{\theta} - \dot{\theta}^2 \theta \right). \tag{3.10}$$

Since load swinging is an unpleasant phenomenon, many methods have been proposed in the literature to eliminate it. A detailed review of methods and publications is available in [5]. Despite the large number of proposed methods, the number of experimentally tested systems is much smaller, and the number of practically implemented methods is even smaller. This is determined by such reasons as insufficient consideration of the physical capabilities of controlling a multi-ton platform in adverse environmental conditions (impossibility or insufficient reliability of the required sensors, especially when weather conditions worsen, increased wear of mechanical and electrical equipment with the complication of the velocity schedule of trolley moving, the possibility of slippage, variability of the required device operation program, etc.). In addition, when using most of the proposed methods, the time for moving the trolley to a given position turns out to be greater than in the system used in previous models, when the trolley accelerates according to a linear law with the maximum acceleration under given conditions. True, in cases where, according to the conditions of application, a very accurate fixation of the direction of movement of the load is required (for example, when loading into hatches), the loss of time for trolley movement is fully or partially compensated by a decrease of the time to dump the load after the trolley stops. Therefore, simple methods belonging to the so-called Input-Shaping are used mostly. One of the simplest ways is to choose the value of the trolley acceleration depending on the length of the rope. If to take in (3.9) the constant value of the trolley acceleration $\ddot{S} = a$ and the length of the rope l and neglect the second term, then the angle θ changes according to the law

$$\theta = \frac{a}{l\omega_0^2}(1 - \cos\omega_0 t), \quad \omega_0 = \sqrt{\frac{g}{l}}. \tag{3.11}$$

The angle θ becomes equal to zero at $t = T = \dfrac{2\pi}{\omega_0}$, and if to choose the acceleration value such that the trolley speeding up ends at $t = T$, that is, $a = v_{max}/T$, then the angle θ will remain equal to zero hereinafter.

If, with the beginning of the movement of the trolley, the lifting of the load begins with the speed v_l, then the calculated value of T can be found based on the average length of the rope $l_m = H - 0.5Tv_l$ that is from the relationship

$$T = \frac{2\pi}{\sqrt{g}} \sqrt{H - 0.5Tv_l} \qquad (3.12)$$

which reduces to a quadratic equation

$$T^2 + 2.01v_l T - 4.02H = 0. \qquad (3.13)$$

The difficulty of applying this method is that before beginning the trolley speed change (acceleration or deceleration), it is necessary to know not only the length of the suspension but also its subsequent movement, that is the mode of operation must be strictly regulated.

Crane1d# models are designed to quickly determine the effect of different systems on load vibrations. In them, the electric motor with a torque control system and a power supply system is replaced by a lag element with an equivalent time constant of 5 ms. Only the speed controller and the motor torque equation are retained. The same simplifications have been applied to some of the models discussed in previous chapters. The display does not show the position of the trolley, but the position of the load, which differs from the position of the trolley by $\Delta S = -l\sin\theta \approx -l\theta$.

Under simulation, the lifting height of the load is 20 m, and the distance of its movement is 90 m.

In the **Crane1d** model, the sequence of events is as follows: lifting a load to a height of 20 m; after a time of $20/1.2 \approx 17.4$ s, when the length of the rope is 15 m, the acceleration of the trolley begins. The acceleration time is assumed to be $= \dfrac{2\pi\sqrt{15}}{\sqrt{9.81}} = 7.8$ s. The deceleration time is the same, since the length of the suspension has not changed. In this case, the deceleration distance is $0.5 \times 4 \times 7.8 = 15.6$ m, so deceleration starts when the trolley has traveled 74.4 m. After the trolley stops, the load is lowered. It can be seen from Figure 3.6 that when the trolley moves at a steady speed and after it stops, there are no swings of the load, and the load drops exactly at the given distance of 90 m. The cycle of operation is equal to 65 s.

In the **Crane1d4** model, the lifting and lowering of the load start simultaneously with the start of the acceleration or deceleration of the trolley, respectively. Based on (3.12), by lifting $T = 10.7$ s, by lowering ($H = 15$ m, $v_l < 0$) $T = 9.06$ s. From Figure 3.7 it can be seen that at $t = 24$ s, the trolley slows down and the load lowers. At $t = 33$ s, the trolley stops at a distance of 90 m, and the lowering of the load continues until $t = 40$ s. During trolley travel at a steady speed and after it stops, the swings of the load are very small.

The method of the crane control based on the use of a double-notch filter proposed in [3] is simulated in the **Crane1d1** model. Figure 3.8 shows the scheme for the fabrication of the speed reference. The staircase speed command (the first step is about 0.25 nominal) and the input lag element serve to reduce the jerk when the command changes, which is observed at the output of the notch filter. The second-order notch

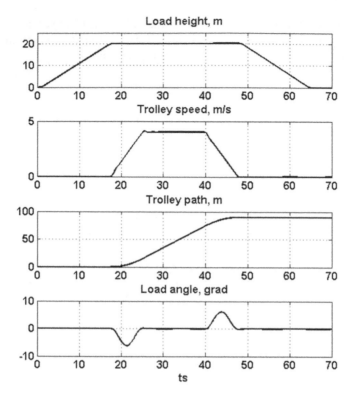

FIGURE 3.6   The movement of the load when lifting and lowering and the movement of the trolley alternate.

filter is implemented as two series-connected first-order filters of a special structure that allow changing their parameters during operation.

Processes with various values of ζ and ω_0 depending on time can be investigated during simulation. As an example, the process is shown in Figure 3.9 when the values of these parameters change from 0.9 to 1.6 and from 0.4 to 1, respectively, when the trolley begins to slow down. One can see that when the load reaches zero level, the trolley stands still already. The load vibrations are small, and when the load reaches zero level, the angle of deviation from the vertical is approximately 1°. The duration of the cycle is 53 s. The disadvantage of this method is the large stopping distance of the trolley and its dependence on the selected values of the filter parameters and the instants of their change.

The system, in which the change of the trolley travel speed is carried out in two steps [6], is simulated in the **Crane1d2** model. It has already been said above that when starting (or braking), the angle θ becomes equal to zero at $t = T = \dfrac{2\pi}{\omega_0}$, and if to choose the acceleration value such that the trolley speeding up ends at $t = T$, that is, $= v_{max}/T$, then the angle θ will remain equal to zero hereafter. In the system under consideration, each step has a value of 0.5 v_{max}, and the acceleration value is equal to the maximum design value a_{max}. In the first stage, after speeding up to the

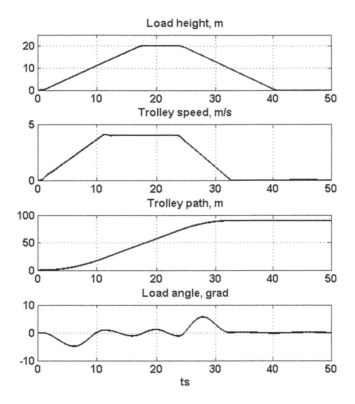

FIGURE 3.7 The movement of the load when lifting and lowering and the movement of the trolley take place simultaneously.

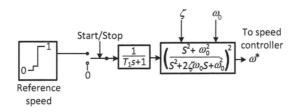

FIGURE 3.8 Scheme for fabrication of the speed reference in the **Crane1d1** model.

speed of 0.5 v_{max}, the acceleration becomes zero, so that the total duration of the first stage is 0.5 T. Then the trolley accelerates to the full speed with the same acceleration (Figure 3.10). Thus, the total acceleration time is $t_a = 0.5 (T + v_{max}/a_{max}) < T$ in the general case. For example, in our case, with $l = 35$ m, $T = 11.9$ s, and the value $t_a = 0.5(11.9 + 4/0.8) = 8.5$ s.

Process is shown in Figure 3.11 when, at first, the load is being lifted to the height of 5 m with the stationary trolley, then the trolley speeds up to a speed of 2 m/s with an acceleration of 0.8 m/s². Since the length of the rope during acceleration is 30 m, then the second acceleration to travel speed of 4 m/s occurs, later $\dfrac{\pi\sqrt{30}}{\sqrt{9.81}} = 5.5$ s

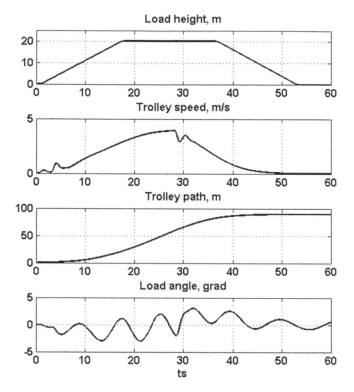

FIGURE 3.9   The movement of the load in the system with the double-notch filter.

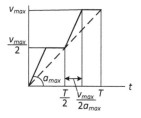

FIGURE 3.10   Two stage speeding up illustration.

after the first one. During the movement of the trolley at half speed, the load begins
to rise to a wanted height (20 m).

Duration of the first stage of the trolley slowing down is $\dfrac{\pi\sqrt{15}}{\sqrt{9.81}} = 3.9\,s$. After the

trolley came to stop at the wanted point (90 m), the load lowering occurs. It is seen that
during trolley travel at a steady speed and under load lowering, the angle of deviation
of the rope from the vertical does not exceed 0.6°. The duration of the cycle is 54 s.

Figure 3.12 shows the process when the load begins to rise at the same time as the
trolley begins to move. It can be seen that when the trolley moves at a steady speed

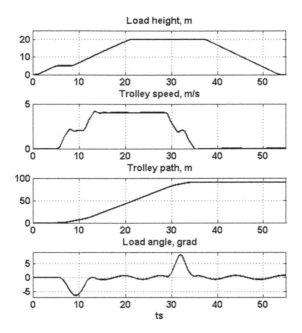

FIGURE 3.11 Load rise and lowering in the system with two-stage speeding up (first operation mode).

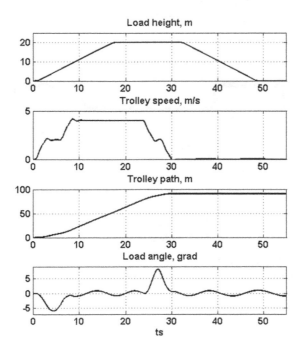

FIGURE 3.12 Load rise and lowering in the system with two-stage speeding up (second operation mode).

and when the load is lowered, the angle of deviation of the rope from the vertical does not exceed 1°. The duration of the cycle decreased to 49 s.

In the **Crane2d**, **Crane2d1**, and **Crane2d2** models, both traveling crane drives are simulated which have a common source in the DC link, as in the **Crane1c** model. Circuits simulating load vibrations and modified control systems from the **Crane1d**, **Crane1d1**, and Crane1d2 models, respectively, have been added to these models. After running the simulation, it can be seen that the processes of lifting and moving the load, the rope angle fluctuations are almost the same as in these latter models, that is as shown in Figures 3.9, 3.10, and 3.12, so the corresponding plots are not shown here. Plots of changes of the hoisting motor torque, of a more loaded trolley motor (for the last figure—for the option when lifting and moving start at the same time) and of the active power consumed or given to the network are shown in Figures 3.13, 3.14, and 3.15 for the models **Crane2d**, **Crane2d1**, and **Crane2d2**, respectively. Reactive power in all cases is close to zero.

3.1.2 CABLE CRANES

In cable cranes, the trolley moves along a carrying rope mounted on supports (Figure 3.16). They are used to transport a load through water obstacles and long structures. The span of the crane can reach 1000 m or more, the load capacity can exceed 50 tons. The movement of the trolley is carried out using cable traction. The lifting and traction winch motors are installed permanently in the towers.

The cable crane will be considered that has a span of 710 m, with a lifting capacity of 25 tons. The lifting height is 90 m, the lifting speed is 96 m/min, the trolley travel

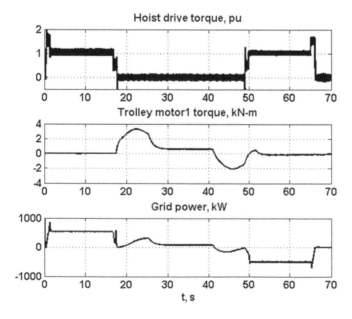

FIGURE 3.13 Traveling crane torques and power under load transportation for the **Crane2d** model.

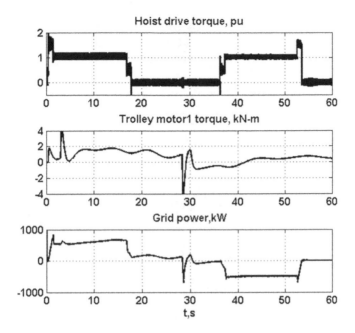

FIGURE 3.14 Traveling crane torques and power under load transportation for the **Crane2d1** model.

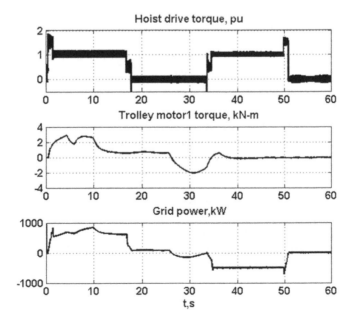

FIGURE 3.15 Traveling crane torques and power under load transportation for the **Crane2d2** model.

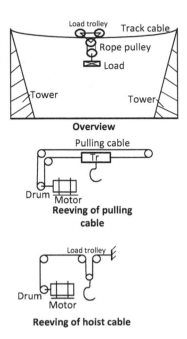

FIGURE 3.16　Overview of the cable crane.

speed is 360 m/min, the diameter of the running wheels is 630 mm, the rope drums are 1200 mm, and the rollers are 320 mm. Acceleration when lifting a load is 0.8 m/s², a trolley acceleration is 1 m/s².

The lifting mechanism is considered at first. The static power is

$$P_l = \frac{25 \times 9.81 \times 96}{60 \times 0.85} = 462 \text{ kW.}$$

Here 0.85 is the mechanism efficiency. A twin pulley block is used. Then the drum rotational speed is

$$N_{dl} = \frac{30 \times 96 \times 2}{\pi \times 60 \times 0.6} = 51 \text{ rpm.}$$

Considering that for the crane under consideration, the duration of lifting the load is much less than the time of its horizontal movement, the IM with a power of 325 kW and with three pole pair ($Z_p = 3$) is taken. IM parameters are: $U_{nom} = 690$ V, $N_{nom} = 992$ rpm, efficiency $= 96.1\%$, cos $\varphi = 0.78$, $I_{nom} = 362$ A, starting current 5.3 pu, no-load current 170 A, rated torque 3130 N-m, maximal torque 2.1 pu, starting torque 0.8 pu, $J_m = 5.7$ kg-m². The motor drives the winch through a two-stage gearbox with $i = 17.2$.

The moment of inertia of the brake disk and the first gearbox pinion which are direct-connected with the IM is taken equal to $J_{ms} = 4.3$ kg-m², the moment of inertia of the drum and four rollers (Figure 3.16) is taken equal to 820 kg-m². By (3.1), taking into account the twin pulley block, we receive

$$J_l = 5.7 + 4.3 + 820/17.2^2 + 0.25 \times 0.25 \times 25 \times 10^3 \times \frac{1.2^2}{17.2^2} = 20.4 \text{ kg-m}^2.$$

Since the lifting speed of 1.6 m/s corresponds to the motor rotational speed of 51 × 17.2 × π/30=91.8 rad/s, the dynamic torque when the load is rising is M_{din}=0.5 × 20.4 × 91.8=936 N-m. Since the static torque is 462/91.8=5.03 kN-m, the total torque is 5966 N-m and is less than the maximum permissible torque that is equal to 2.1 × 3130=6573 N-m.

Since the static torque ratio is 5030/3130=1.6 and the load lifting time is about 1 min, and in consideration that the motor permits an overload of 1.5 pu during 2 min, the given static torque ratio may be considered acceptable. If to take the minimum realized trolley travel equal to 0.8 × 710=568 m, then the time of its movement will be 568/360=1.58 min. If to designate k_i as the load factor in the i-th interval of the cycle and take the load torque when lifting an empty hook equal to 0.3 from the torque of full load (i.e., k_i=0.3 × 1.6=0.48 in this section), then it can be accepted approximately: loading, t_1=1 min, k_1=0; lifting t_2=1 min, k_2=1.6; movement t_3=1 min (assuming that the lifting and moving of the load are partially performed simultaneously), k_3=0; lowering t_4=1 min, k_4=1.6; unloading t_5=0.5 min, k_5=0; lifting t_6=1 min, k_6=0.48; movement t_7=1 min, k_7=0; lowering t_8=1 min, k_8=0.48. Therefore, motor rms load is

$$k_{eq} = \sqrt{\frac{2 \times 1.6^2 + 2 \times 0.48^2}{7.5}} = 0.86,$$

that is, less as 1.

Consider the choice of the motor for the winch which moves trolley. Circumferential force on the traction drum is

$$F_T = W_T + W_l \tag{3.14}$$

where W_T is a trolley tractive resistance, W_l is losses on the guide blocks of the traction rope.

$$W_T = Q(w\cos\gamma + \sin\gamma) \tag{3.15}$$

where Q is the weight of the trolley with a load, w is the coefficient of the specific resistance to movement (it is taken as w=0.02), γ is the angle of inclination to the horizon of the trajectory of the trolley movement which is a variable depending on the current location of the trolley.

The quantity can be determined as

$$W_l = T \sum_i^n (1 - \eta_i) \tag{3.16}$$

where T is the rope tension, η_i is the efficiency of the guide blocks, $n = 3$ in Figure 3.16. The quantity $T = T_0 + F_T$ where T_0 is the initial, so-called mounting tension, which can be estimated as

$$T_0 = \frac{q l_p^2}{8f},$$ (3.17)

q is the linear gravity force (accept in advance 25 N/m), l_p is the distance between supports (60 m), f is a sag, $f \approx 0.008 l_p$, so $T_0 \approx 25$ kN.

Provided that the heights of fixing the carrier cable are equal

$$\gamma = arctg\left[\frac{l_x - 0.5l}{T_{01}}\left(q + \frac{Q}{l} \right) \times 9.81 \right]$$ (3.18)

Here T_{01} is the tension of the carrying rope, q is its linear weight, l is the span of the crane, l_x is the current position of the trolley, counted from the initial support. When traveling on the first half of the path, the presence of a slope reduces the resistance to movement, and on the second half, it increases. It is possible to accept for the considered crane $T_{01} = 1775$ kN with $q = 27.7$ kg/m. If the efficiency of the guide block is taken equal to 0.96 then

$$F_T = \frac{W_T + 0.12T_0}{0.88}$$ (3.19)

The tractive force depending on the trolley location is shown in Figure 3.17 if the mass of the trolley is equal to 20% of the mass of the load.

Usually, the loaded trolley does not reach the end positions. We assume that it does not reach 20 m to each tower. Then the maximum value of the traction force is 56 kN. Suppose that such an effort can be realized when the motor is overloaded by 1.6 times. Then the rated motor torque is calculated for $F_T = 35$ kN, which corresponds to

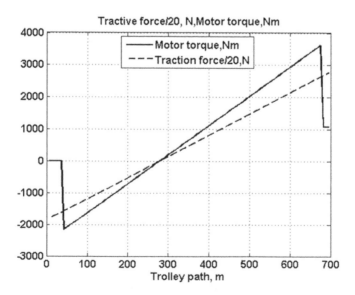

FIGURE 3.17   Dependence of the tractive force on the trolley location for the cable crane.

a drum torque of 21 kN-m and a power of 210 kW. The drum rotation speed is 10 rad/s or 95.5 rpm. Taking the efficiency of the winch equal to 0.9, choose the IM motor with the power of 250 kW and $Z_p = 3$. IM parameters are: $U_{nom} = 690$ V, $N_{nom} = 991$ rpm, efficiency $= 95.9\%$, cos $\varphi = 0.78$, $I_{nom} = 279$ A, starting current 5.2 pu, no-load current 132 A, rated torque 2408 N-m, maximal torque 2.1 pu, starting torque 0.8 pu, $J_m = 4.8$ kg-m^2. The motor drives the winch through a two-stage gearbox with $i = 9.73$.

The moment of inertia of the brake disk and the first gearbox pinion which are direct-connected with the IM is taken equal to $J_{ms} = 4.3$ kg-m^2, the moment of inertia of the drum and three rollers (Figure 3.16) is taken equal to 790 kg-m^2. By (3.1), we receive

$$J_l = 4.8 + 4.3 + 790/9.73^2 + 0.25 \times 3 \times 10^4 \times \frac{1.2^2}{9.73^2} = 131.5 \text{ kg-m}^2.$$

Since the travel speed of 6 m/s corresponds to the motor rotational speed of 97.3 rad/s, the dynamic torque is $M_{din} = 131.5 \times 97.3/6 = 2132$ N-m.

When modeling the trolley electric drive, the elastic connection between the drum and the trolley through a long cable is taken into account. Relative value of elastic deformation under tension T

$$\frac{\Delta L}{L} = \frac{T}{K} \tag{3.20}$$

where the rope linear rigidity $K = ES$, E is the elastic modulus of the rope, N/mm^2, S is the cross-sectional area of the wires, mm^2. The value of E differs from the value of the elastic modulus of a rigid rod due to the large number of wires wrapping around each other. In addition, the value of E depends on the tension, which, as mentioned above, is a sum of the mounting tension and the force of the trolley movement. Nevertheless, for simplicity, we will assume the value of E constant, equal to 1.6×10^5 N/mm^2 for the traction rope. Assuming the rope safety factor for breaking equal to 5 and taking into account that the maximum tension value is $(25 + 56)$ kN, a rope with a diameter of 28 mm is selected, $q = 29$ N/m, $S = 298$ mm^2. Then

$$T = \frac{1.6 \times 10^5 \times 298}{l} (l_1 - l_2) = \frac{4.8 \times 10^7}{l} (l_1 - l_2) \tag{3.21}$$

l is the length of the rope from the point of its leaving the drum to the trolley, we assume that l varies from 120 to 780 m, l_1 is the linear movement of the leaving point of the rope from the drum, l_2 is the movement of the trolley. The behavior of the system is described by the equations

$$J_1 \frac{d\omega_m}{dt} = M_m - \frac{D_r}{2i} T \tag{3.22}$$

where J_1 is the total moment of inertia of the motor and the drum with a diameter D_r reduced to the motor shaft, ω_m, M_m are the motor rotational speed and the torque, respectively.

$$\frac{dl_1}{dt} = \frac{D_r}{2i}\omega_m \tag{3.23}$$

$$m\frac{dV}{dt} = T - F_T \tag{3.24}$$

$$\frac{dl_2}{dt} = V \tag{3.25}$$

The dynamics of the change of T has significant damping properties caused by the slip of the rope wraps along the drum and guide rollers, the mutual slip of the rope wraps, and so on. Therefore, instead of (3.21), the relationship (3.26) is used:

$$T = \frac{2\times10^7}{l}\left(1 + k_d\frac{d}{dt}\right)(l_1 - l_2). \tag{3.26}$$

There is very little actual data on the possible damping factors for ropes used in hoisting machines. For example, in [7], data on the damping coefficients of lifting ropes are given. But they differ in design from the considered traction ropes. Therefore, the simulation results obtained below should be regarded as indicative.

The **Crane4** model is designed to familiarize with the features that arise when taking into account the elasticity of the traction rope. The model contains two simplified models of the trolley electric drive: **Trolley drive** and **Trolley drive1**; the first takes into account the elasticity of the traction rope, and in the second model, motor, drum, and trolley are considered a single concentrated system. The electric drive in these models is modeled in a simplified way, as in the **Crane1d2** model. The model provides the ability to simulate modes both with a constant rope length and with a rope length that varies according to a linear law. To reduce the fluctuations of the load hanging on the rope, a system with a speed setting in two steps was used, as in the **Crane1d2** model.

Taking into account the specificity of the application of the cranes under consideration, when the load must be carried over high structures (dams under construction and nuclear power plants, river vessels, etc.), it can be assumed that the main mode of operation will consist of sequential operations: lifting the load to a given height, moving the trolley with the load at a constant height, lowering the load with a stationary trolley.

According to (3.11), (3.12), when the rope length is 30 m, the first step duration is $t_1 = \pi\sqrt{30}/\sqrt{9.81} = 5.5$ s; the same is the duration of the first step by slowing down when during the trolley travel, the rope length does not change. The processes are shown in Figure 3.18 when the trolley moves with a fixed load at such a height that the length of the rope is 30 m. For the system that takes into account the elasticity of the rope, the value $k_d = 0.1$. It can be seen that for the rigid system, the angle of deflection of the load when the trolley moves at a constant speed is equal to zero, while when elasticity is taken into account, vibrations of the load with an amplitude of 0.3° are observed, and at the moment the trolley stops, the angle is equal to −1.8°.

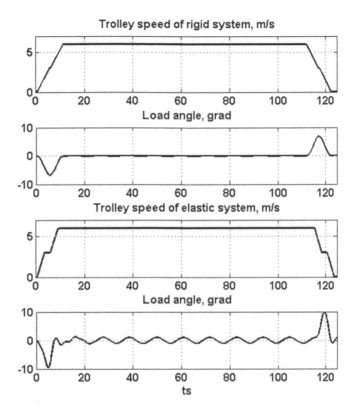

FIGURE 3.18   Trolley travel and the load angle vibration in the cable crane with and without taking into account the rope elasticity.

At this, the travel of the loaded trolley occurs from the tower with the winch. If to perform a simulation when the loaded trolley moves to the tower with the winch, one can see load vibrations with the angle amplitude of 1.2°, and the angle is −1.2° at the instant when the trolley stops.

The **Crane3** model simulates the electric drive of the trolley, taking into account the fluctuations of the suspended load and the changing length of the rope. An indirect vector control system, described, for example, in [8], is used to control the motor. The input of the control system receives signals for setting the torque T^* (output of the speed controller) and the rotor flux Ψ_r^* (this value is constant up to the nominal speed of rotation of the IM ω_m and decreases inversely with ω_m at $\omega_m > \omega_{nom}$).

In the control system, the set points of the currents I_d^* and I_q^* are fabricated.

The first quantity is generated by the PI controller of the rotor flux linkage, at the input of which the reference value Ψ_r^* and the calculated value Ψ_r (see below) are compared. The set point of the I_q^* is calculated from relation (3.27):

$$I_q^* = \frac{2}{3} \frac{L_r}{L_m} \frac{T_e^*}{Z_p \Psi_r}$$

(3.27)

Here $L_r = L_{lr} + L_m$, L_{lr} is the leakage inductance of the rotor winding; L_m is the IM main inductance. Computed values of I_d^*, I_q^* are transformed in the set points for the phase currents I_{abc}^* in the stationary reference frame, supposing that the angle θ of the vector $\boldsymbol{\Psi}_r$ is known.

To calculate $\boldsymbol{\Psi}_r$ and θ, the measured IM phase currents are used. They are converted in the quantity I_d, I_q by the conversion block **ABC-DQ**. The value

$$\Psi_r = \frac{L_m I_d}{1 + s T_r} \tag{3.28}$$

where $T_r = L_r / R_r$, s is the Laplace transformation symbol; the value of the rotational speed of the rotor linkage vector in the stator reference frame ω_s is equal to the sum of the mechanical speed $\omega_r = Z_p \, \omega_m$ and the slip frequency $\Delta \omega$:

$$\Delta \omega = R_r \frac{L_m}{L_r} \frac{I_q}{\Psi_r} \tag{3.29}$$

The integral of ω_s is an estimation of θ.

The library contains subsystem **AC3** which models the considered electric drive control system. However, in the library model, the inverter is powered by a diode rectifier which does not allow the recovery of electrical energy. Initially, when developing this model, an attempt was made to make the necessary changes in **AC3**, after breaking the connection with the library model, but it turned out that the model, which was modified when using one version of MATLAB, may not work when using another version. Therefore, the model of the IM indirect vector control with the possibility to use an input active rectifier was developed. In this model, subsystems from the library model were used to the maximum extent possible.

In the **Crane3** model, the elasticity of the traction rope is not taken into account. The speed control of the trolley is carried out in two steps, as in the model **Crane1d2**. Figure 3.19 shows the process when the acceleration of the trolley begins with the load already lifted to a height of 60 m (rope length 30 m). After the acceleration of the trolley, further lifting of the load takes place (the length of the rope is 10 m). Under approach to a place of unloading, the load is lowered by 20 m, so that the braking of the trolley occurs with a rope length of 30 m.

From a comparison of Figure 3.19 with Figure 3.18, it can be seen that a full account of the dynamics of the electric drive of the trolley leads to some increase in the load angle deflection of the vertical when its speed changes, but the general nature of the change in the angle is preserved.

The **Crane31** model takes into account the elasticity of the traction rope, the damping coefficient is taken equal to 0.05. Figure 3.20 shows the same process as Figure 3.19 when the trolley with a load moves from a tower with a winch. One can see an increase in vibrations when the trolley stops, however, at the moment of its stop (119 s), the deviation angle is less than 2°. Figure 3.21 shows the process when the trolley with a load moves to the tower with a winch. It can be seen that, in contrast to the previous case, increased vibrations are observed at the beginning of the movement, when the length of the traction rope is maximum.

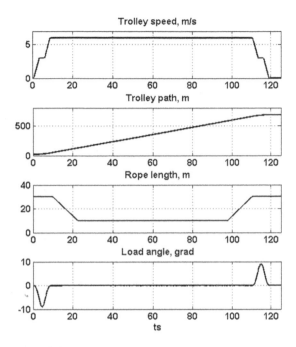

FIGURE 3.19 Trolley travel and the load angle vibration in the cable crane without taking into account the rope elasticity, with taking into account dynamics of the electric drive.

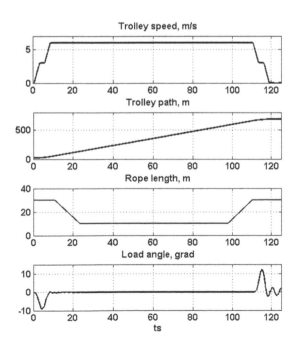

FIGURE 3.20 Trolley travel and the load angle vibration in the cable crane with the elastic rope, movement from a tower with a winch.

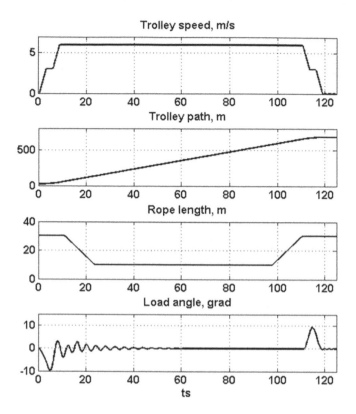

FIGURE 3.21 Trolley travel and the load angle vibration in the cable crane with the elastic rope, movement to a tower with a winch.

The **Crane32** model simulates the hoisting electric drive. In accordance with the accepted efficiency of the mechanism, the torque of the load is 85% of the total torque and is always directed downward, and the torque of losses is 15% and is directed against the direction of rotation of the drum. When the motor is stopped, the brake is applied and takes over the torque created by the hanging load. The process is shown in Figure 3.22 when the load initially rises to a height of 30 m, then at $t=25$ s to a height of 35 m, and at $t=40$ s a command is given to lower the load. Recall that the path traveled by the load during acceleration or deceleration with a speed of 1.6 m/s and an acceleration of 0.8 m/s² is 1.6 m.

The **Crane33a** model simulates the joint operation of both mechanisms. Both drives are supplied by the common active front rectifier with a DC voltage of 1200 V. The rectifier is powered from the network 6.3 kV with the help of the transformer 6300/750 V and reactor. The following mode is modeled: (1) Lifting the load to a height of 60 m (rope length $l=30$ m); to reduce the simulation time, which turns out to be quite large, the initial height is taken equal to 30 m. (2) Trolley acceleration in two steps. (3) After the trolley accelerates to a speed of 6 m/s, the hoist is turned on to further reduce l to 10 m. (4) When approaching the final point ($S_e=690$ m), at $S=580$ m, a command is given to lower the load to a height of 60 m. (5) At $S=665$ m,

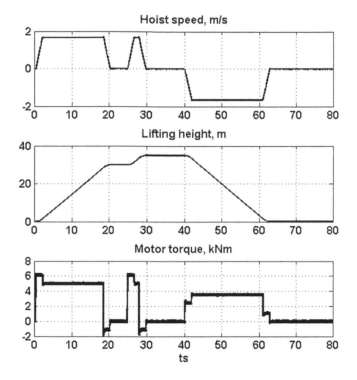

FIGURE 3.22 Lifting and lowering of the load.

the trolley slowing down process begins in two steps. (6) After the trolley stops, the load is lowered; to reduce the simulation time, the final height is also assumed to be 30 m. In this model, the elasticity of the traction rope is not taken into account.

Some results of simulation are shown in Figures 3.23 and 3.24. It is seen that the angle of the load deviation from a vertical is getting zero after the end of the trolley speeding up and when the trolley stops. Maximum of the active power consumed from the grid is not more than 600 kW, the reactive power is by an order less. THD in the grid current, when it reaches its maximal amplitude, is about 3.5%.

3.2 MINE HOIST

Modern mine hoisting installations can have a lifting height of up to 1.5 km, have a carrying capacity of up to 50 tons, move at a speed of up to 20 m/s, and represent a complex electromechanical system consisting of a number of concentrated masses (motors, drums, pulleys, lifted conveyances-cage or skip with load, counterweights) connected by elastic links (shafts, ropes). There are drum-type lifting installations, in which the rope is attached to the drum and wound around it, and with a friction pulley, in which the ropes (usually several) move due to friction forces, being pressed against the pulley by the weight of the load and counterweight.

Figure 3.25 shows the arrangement of these two types of lifting installations, where it is designated: (1) winch drum with a motor, (2) guide pulleys, (3) a conveyance with

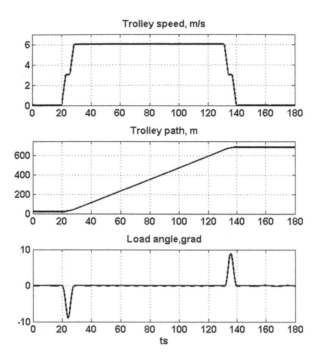

FIGURE 3.23 Joint operation of the trolley and hoist; trolley speed and travel, load angle.

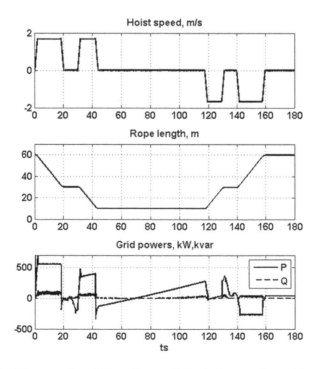

FIGURE 3.24 Joint operation of the trolley and hoist; hoist operation, grid power.

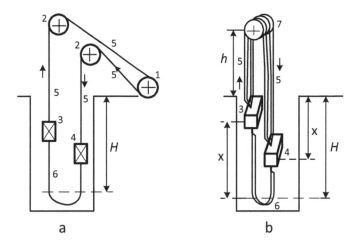

FIGURE 3.25   Arrangements of the lifting installations.

a load, (4) a counterweight or a second conveyance, (5) head lifting ropes, (6) closing, or tail ropes, (7) friction pulley.

Firstly, modeling of the hoist with the friction pulley is under consideration. The parameters are: a cage with a mass of $G_k = 27\,t$ can lift a load up to $G_l = 30\,t$, so that the maximum weight to be lifted is $G_u = 57\,t$. The weight of the counterweight is selected according to the formula $G_b = G_k + 0.5G_l = 42\,t$. The mine depth is $H = 920\,m$. The head-frame height $h = 80\,m$. The number of head ropes is 4 with a linear mass $p = 8.38$ kg/m and a diameter $d_k = 38\,mm$. The number of tail ropes is 2 with a linear mass $q = 17.25$ kg/m and a diameter $d_{tk} = 64\,mm$. The difference between the total linear masses of the tail and head ropes $\Delta = 2 \times 17.25 - 4 \times 8.38 = 0.98$ kg/m. The diameter of the rope driving friction pulley is assumed to be $D_{sh} = 5\,m$.

When the cage is moving, additional resistances arise due to friction forces in the guide shoes of the moving cage, aerodynamic resistance to the movement of the cage in the limited space of the mine shaft, friction forces in the bearing supports of the rope-carrying equipment, and so on. The exact calculation of these resistances is complicated, therefore it is assumed that the resistance value is equal to $k'G_l$, $k' = 0.15 - 0.2$. With this in mind, the static tension of the rising branch when the cage is raised to a height x (Figure 3.25) is equal to

$$S_{up} = g\left[(1 + k'/2)G_l + G_k + 4p(H + h - x) + 2qx\right] \tag{3.30}$$

and the lowering branch

$$S_d = g\left[0.5(1 - k')G_l + G_k + 4p(h + x) + 2q(H - x)\right]. \tag{3.31}$$

Then the static force of the hoisting system is

$$F = S_{up} - S_d = g\left[(0.5 + k')G_l + (4p - 2q)(H - 2x)\right]. \tag{3.32}$$

Thus, taking further $k'=0.2$, at the lower position of the cage (further, for definitiveness, the term cage is used instead of conveyance) $F_1 = 9.81 \times (0.7 \times 30 - 0.98 \times 920 \times 10^{-3}) = 197.2$ kN, at the upper position 214.9 kN.

A synchronous motor with $Z_p = 10$ and with a rated frequency of 10 Hz is accepted for the drive of the friction pulley, that is, as if a reduced version of the motor used in the first chapter for the plate mill. The drive is gearless. Under the rated rotational speed of 6.28 rad/s, the load travel speed is $6.28 \times 5/2 = 15.7$ m/s. The maximum static torque is $214.9 \times 2.5 = 537$ kN-m, the average lifting power $P_{st} = 0.5 \times 537 \times (1 + 197.2/214.9) \times 6.28 = 3233$ kW.

The motor 4000 kW, 3150 V, 60 rpm, $J_m = 50,000$ kg-m² is chosen. The possibility to receive an acceleration of 1 m/s² is checked when the moment of inertia of the friction pulley is 156,000 kg-m².

The total reduced (equivalent) mass of the hoisting system is $M_{ekv} = 27 + 30 + 42 + (2 \times 17.25 \times 920 + 4 \times 8.38 \times 1080) \times 10^{-3} + (50 + 156)*4/5^2 = 200$ t. By acceleration of 1 m/s², the dynamic force is 200 kN and the pulley torque is 500 kN-m; the motor-rated torque is $M_{nom} = 4000/2\pi = 637$ kN-m. Therefore, $M_{max}/M_{nom} = (537 + 500)/637 = 1.63$.

Three-step velocity diagram is considered, neglecting a draw time before the complete stop.

The speeding up time is 15.7 s, the path is $S_1 = 15.7 \times 15.7/2 = 123$ m, motor torque in pu is $M_1^* = 1.63$. The slowdown time is the same, the torque $M_3^* = \dfrac{-500 + 537}{637} = 0.06 \approx 0$. The time of moving with a steady speed is $t_2 = (920 - 2*123)/15.7 = 42.9$ s, $M_2^* = \dfrac{537}{637} = 0.84$. Pause duration is taken equal to 15 s, then $M_{rms}^* = \sqrt{\dfrac{1.63^2 \times 15.7 + 0.84^2 \times 42.9}{15.7 \times 2 + 42.9 + 15}} = 0.9$.

A special feature of a dynamic operation of lifting systems with friction pulleys is that in forced start and stop conditions, there is a danger of the rope slipping along the surface of the rope-carrying body. The need to exclude such a slip of the rope is an additional factor limiting the magnitude of the starting acceleration, as well as the deceleration of the lifting system when it stops.

Under acceleration of the lifting system, the condition of the absence of rope slip is described as follows:

$$S_{up} < S_d e^{f\alpha} \tag{3.33}$$

(see formula (2.11)). Or

$$S_{up} - S_d < S_d\left(e^{f\alpha} - 1\right). \tag{3.34}$$

The wrap angle α is taken equal to π ($\alpha = \pi$), the friction coefficient $f = 0.2$. At the beginning of the lifting, the mass of the rising parts is 90.5 tons, and of the lowering parts is 76.4 tons. By (3.30), (3.31) $S_{up} = 917.4$ kN, $S_d = 720.3$ kN. Therefore, the tension of the rising rope branch, at the lifting beginning with the acceleration of 1 m/s² is $(917.4 + 90.5 \times 1) = 1007.9$ kN, and the lowering rope branch $(720.3 - 76.4 \times 1) = 643.9$ kN, from which $1007.9 - 643.9 = 364 < 643.9 \times (e^{0.628} - 1) = 563$.

The static tension of the rising rope branch when the load is in the upper position is $S_{up}=926.3$ kN, and the lowering rope branch $S_d=711.4$ kN. At the lifting end, the mass of the rising parts is 91.4 tons, and the lowering ones is 75.5 tons. Thus, the tension of the rising rope branch at the lifting end, when the load is slowing down with the deceleration of 1 m/s², is $926.3-91.4=834.9$ kN, and that for the lowering rope branch is $711.4+75.5=786.9$ kN. Thus, the condition (3.34) is satisfied too. It is worth noting that if it turned out $S_{up}<S_d$ during deceleration at the lifting end (which could take place with an increase of deceleration), then instead of (3.34), the condition

$$S_d - S_{up} < S_{up}\left(e^{f\alpha} - 1\right) \tag{3.35}$$

should be fulfilled.

The **Mine1a** model simulates the hoisting drive with a friction pulley and the above parameters. The entire system is considered single-mass. The electric motor and control system are slightly modified copies of the **Plate3** model, with a threefold reduction in power. Motor static torque

$$T_r = g\left\{(0.5+k')\times 30 - \left[0.98\times(920-2x)\right]\times 10^{-3}\right\}\times 2.5 = 493 + 0.048x \text{ kN-m.}$$

To reduce the oscillations that occur when the speed of movement of the load changes and are associated with the presence of elastic links in the system, a limitation on the rate of acceleration increase, the so-called jerk, is introduced when the speed setting is changed. The simplest way to limit a jerk is to set an element with a first-order transfer function at the output of the rate limiter (Figure 3.26). Speed reference, when the setting value W_{ref} changes as a step, changes as

$$W^* = a\left(Te^{-t/T} + t - T\right) \tag{3.36}$$

The real acceleration $a^* = a\left(1-e^{-t/T}\right)$ and the jerk $a' = \dfrac{a}{T}e^{-t/T}$.

Using this relationship, the time of speeding up to the wanted speed t_w can be found. For example, at $W_{ref}=15.7$ m/s, $a=1$ m/s², $T=2$ s, $t_w \approx 17.7$ s (i.e., under given parameters, $t=W_{ref}/a+T$, because of the first term in (3.36) is about zero). The distance traveled during acceleration (or deceleration) is

$$S = a\left(\frac{t^2}{2} - Tt + T^2(1-e^{-t/T})\right) \tag{3.37}$$

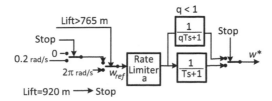

FIGURE 3.26 Block-diagram of the speed setting with jerk limitation.

In our case, $S = 125.2$ m. Figure 3.27 shows plots of changes in speed, travel, torque, and motor power when lifting a load to the height of 920 m. To obtain wanted stop accuracy, deceleration begins at a slightly lower lift height than it follows from the above calculation: 765 m instead of 794.8 m, and a slowdown does not occur to zero speed, but to a speed of 0.5 m/s, the so-called drawing speed. And only at a position of 920 m there is a sharp decrease of the speed to zero with the switch-off or decrease of the time constant of the element that provides the limitation of the jerk (Figure 3.26). It can be seen that the specified accuracy is provided.

Quantities from (3.34) are computed also in the model

$$\sigma = \frac{S_{min}\langle S_{up}, S_d \rangle \times \left(e^{f\alpha} - 1 \right)}{\left| S_{up} - S_d \right|} \tag{3.38}$$

where σ is an antiskid safety coefficient that must be not less than 1.25. The plot of these quantities is given in Figure 3.28.

Take attention that the controller of the reactive power consumed by the motor is included in the system, which affects the setting of the motor current component I_M. Due to this, the reactive power is close to zero (Figure 3.27). Note that without such a controller, the reactive power is quite significant and, for example, at $t = 19$ s it reaches 0.55 of the rated power.

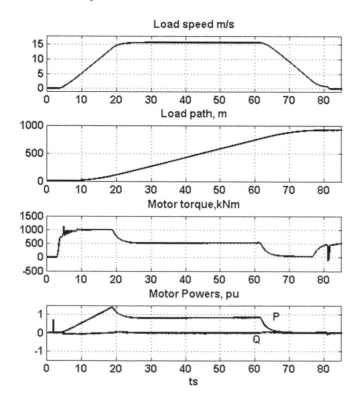

FIGURE 3.27 Processes of the load lifting in the **Mine1a** model.

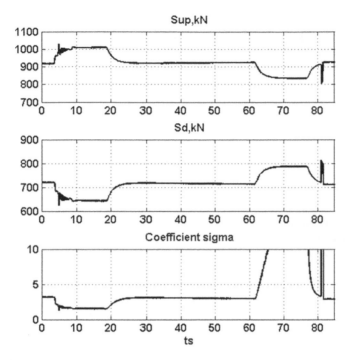

FIGURE 3.28 Rope tension under load lifting in the **Mine1a** model.

The considered models do not take into account the elastic properties of the ropes, which, with their large length, have a significant impact on the operation of the mine hoist. The elastic properties of the rope are characterized by the coefficient of longitudinal stiffness of the rope, which is determined by the formula

$$c_y = \frac{EnF}{L} \qquad (3.39)$$

where E is the modulus of elasticity of the rope, F is the area of the wires in the rope, n is the number of parallel ropes, L is the length of the rope

Strictly speaking, the mass and deformation characteristics of the rope are distributed along its length, however, when this phenomenon is taken into account, the analysis of the system is very difficult, therefore, with sufficient accuracy for practice, the representation of the rope in the form of a weightless thread, with elasticity and damping properties, is used, and the inertial mass of the rope is taken into account by adding 1/3 of the mass of the rope to the masses of the elements that this rope connects [9].

In what follows, the simulated system is considered a three-mass one: motor and friction pulley, cage, and counterweight; only the elasticity of the head ropes is taken into account. The equations of the system have the form.

Motor and friction pulley:

$$m_{me} \frac{dv}{dt} = \frac{T_m}{R_r} - F_{el1} - F_{el2} - F \qquad (3.40)$$

The equivalent reduced mass of the motor and friction pulley, taking 1/3 of the mass of the head rope, is $m_{me} = (50 + 156)/2.5^2 + 1080 \times 4 \times 8.38 \times 10^{-3}/3 = 45.1$ t. T_m is the motor torque, R_r is the friction pulley radius (2.5 m), F_{el1} and F_{el2} are elastic forces of the loaded and idle branches of the head ropes, and F is the static force of the hoisting system which is determined by (3.32) with $k' = 0$.

Loaded conveyance (cage):

$$m_1 \frac{dv_1}{dt} = F_{el1} - Fc. \tag{3.41}$$

Here m_1 is the reduced mass of the loaded cage:

$$m_1 = 57 + (1000 - x) \times 4 \times 8.38 \times 10^{-3}/3 + 2 \times 17.25 \times 10^{-3}x = 68.2 + 23.3 \times 10^{-3}x \text{ t.}$$

Assuming that the forces of resistance to the movement of the loaded cage and counterweight are equal and constant, we have $F_c = 0.5k'Q_l = 30$ kN.

Counterweight:

$$m_2 \frac{dv_2}{dt} = F_{el2} - Fc. \tag{3.42}$$

Here m_2 is the reduced mass of the counterweight:

$$m_2 = 42 + (80 + x) \times 4 \times 8.38 \times 10^{-3}/3 + 2 \times 17.25 \times 10^{-3}(H - x) = 74.6 - 23.3 \times 10^{-3}x \text{ t.}$$

Elastic forces are determined as:

$$F_{el1} = c_{y1}(S - S_1) + d_{y1}(v - v_1) \tag{3.43}$$

$$F_{el2} = c_{y2}(S - S_2) + d_{y2}(v - v_2) \tag{3.44}$$

where S, S_1, S_2 are the displacements which correspond to the speeds v, v_1, and v_2. Here d_{y1} and d_{y2} are damping factors which are conveniently determined in terms of the logarithmic decrement δ. Using the relations given in [9], one can show that

$$\frac{d_y}{c_y} = \frac{\delta}{\pi \omega_r} \tag{3.45}$$

where ω_r is a resonance frequency,

$$\omega_{r1} = \sqrt{c_{y1} \frac{m_{me} + m_1}{m_{me} m_1}}, \quad \omega_{r2} = \sqrt{c_{y2} \frac{m_{me} + m_2}{m_{me} m_2}} \tag{3.46}$$

For the simulated system, the modulus of elasticity $E = 1.2 \times 10^5$ N/mm², the cross-sectional area is 920.6 mm², $n = 4$. Then the linear stiffness of the rope $K = 4 \times 920.6 \times 1.2 \times 10^5 = 4.4 \times 10^8$ N, and in accordance with (3.39)

$$c_{y1} = 4.4 \times \frac{10^8}{1000 - x}, c_{y2} = 4.4 \times \frac{10^8}{80 + x} \tag{3.47}$$

Such a system is simulated in the **mine2a** model. The motor inertia constant H_m is equal in the case to

$$H_m = \frac{206 \times 10^3 \times 6.28^2}{2 \times 4 \times 10^6} = 1.02 \, s$$

The same process as in Figures 3.27 and 3.28 is shown in Figures 3.29 and 3.30, but taking into account the elasticity of the ropes. Figure 3.30 shows the plots of the forces of a rope branch rising with a load and a branch descending with a counter-weight, as well as the resulting load-lifting force. A noticeable increase of vibrations is seen, especially when the cage is stopped. At the very end of slowdown, almost at zero speed, slippage of the ropes along the friction pulley can take place. Significant fluctuations in the rope tension can be seen. However, due to the large mass, the cage stops in a predetermined position without vibrations. This phenomenon is typical in the systems under consideration [9]. Therefore, it is reasonable to simulate the process of safety stop of a loaded cage.

It is assumed in the **mine2b**, **mine2b1** models that at $t = 25$ s, when the cage has been raised by 190 m, a signal is given for a safety brake. In this case, the motor loses power and a signal is given to apply the safety brake. The latter has a response delay of 0.2 s, and the braking force increases exponentially with a time constant of 0.3 s,

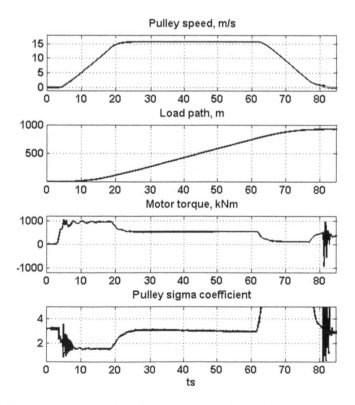

FIGURE 3.29 Processes of the load lifting in the **Mine2a** model.

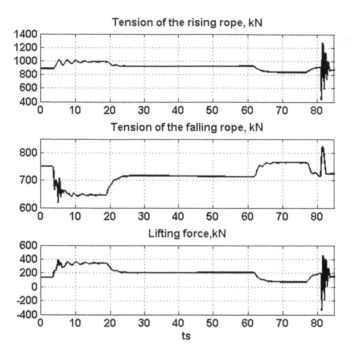

FIGURE 3.30 Rope forces under load lifting in the **Mine2a** model.

so that the brake response time can be estimated at 0.8–1 s. It is assumed that the maximum torque generated by the brake is 1000 kN-m (equivalent force 400 kN).

The **mine2b1** model is considered first. It does not take into account the elasticity of the ropes. The load-lifting force is calculated according to (3.32), but with separate calculation of S_{up}, S_d, and when calculating these values, dynamic components are taken into account, using the actual values of the machine acceleration. These calculations are performed in the **Frict_coeff** subsystem. Figure 3.31 shows plots of the change of the load travel speed, motor torque, and the coefficient σ. It is seen that the system stops for 5 s, that is, the deceleration is 3 m/s². The condition of the absence of the rope slippage along the friction pulley is fulfilled with a margin.

The **mine2b2** model simulates safety braking when lowering a partially loaded cage (approximately 30%). In this case, the losses increase the tension S_{ul} of the unloaded ascending rope and decrease the tension S_l of the partially loaded descending rope. Before the motion starts, the load is held by the brake, which is released 0.3 s after the given speed reference. It can be seen from Figure 3.32, after the beginning of the lowering of the load, the tension of the loaded branch of the rope drops, and the unloaded one increases. At that, the tension difference is 240 kN that corresponds to the motor torque of 600 kN-m. The dynamic torque of the motor itself with a friction pulley is $T_m = \dfrac{2 \times 1.02}{15.7} \times \dfrac{4000}{6.28} = 83$ kN-m, that is, the total torque of the motor during acceleration is approximately 700 kN which can be seen on the **Scope_Drive** oscilloscope. After the start of the safety braking, the motor stops in 6 s, that is,

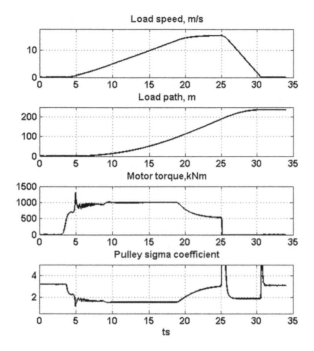

FIGURE 3.31 Safety braking of the rising cage.

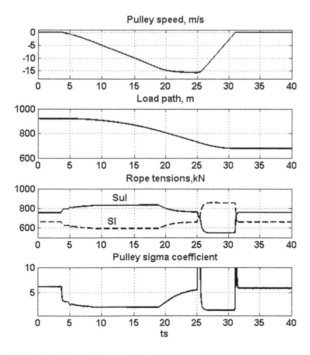

FIGURE 3.32 Safety braking of the lowering cage.

deceleration is 2.6 m/s². The condition of the absence of rope slippage along the friction pulley is fulfilled with a margin.

The conditions change when the elasticity of the head rope is taken into account, as is done in the **mine2b** model. The brake torque is 500 kN-m. Figure 3.33 shows that the motor stops in 4.3 s, that is, deceleration is equal to 3.65 m/s² (according to existing requirements, it should not exceed 5 m/s²). It can be seen from Figure 3.34 that during braking, rope tension vibrations occur with different frequencies (for an unloaded, descending rope, with a higher frequency, since the length of the rope branch is shorter). These vibrations can cause the ropes to slide over the surface of the friction pulley. When braking, the pulley brakes smoothly, and the load and counterweight with an overshoot of 25–30 cm.

In the following models, a two-skip single-rope drum hoisting installation is modeled (Figure 3.25) with the following data [10]: the depth of the mine $H=740$ m, additional movements of the skips for unloading and loading $h_1=h_2=22.5$ m, the total length of one head rope, taking into account rope strings between the drum and the guide pulley and friction wraps, 970 m, skip load capacity $G_l=9.3$ t; the skip empty mass $G_s=8.9$ t, so that the total mass of the skip $G_u=18.2$ t; skip lifting height under unloading $h_p=2.17$ m, maximum speed of movement of the load 10 m/s, acceleration 1 m/s², linear mass of the rope $p=9.11$ kg/m, its diameter 50 mm, head and tail ropes of the same type, drum diameter $D_d=5$ m, moment of inertia $J_d=1.4 \times 10^5$ kg-m², guide pulley diameter $D_t=4.95$ m with moment of inertia $J_t=1.8 \times 10^4$ kg-m².

Since the ropes are completely symmetrical, the static tension does not depend on the position of the skips and is equal for the rope branch with a load

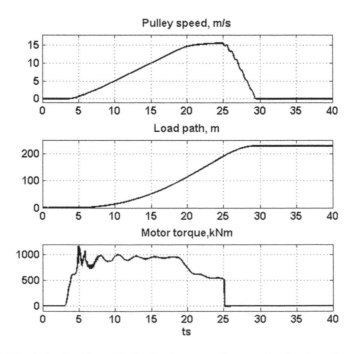

FIGURE 3.33 Safety braking with the elastic rope, pulley speed and path, and motor torque.

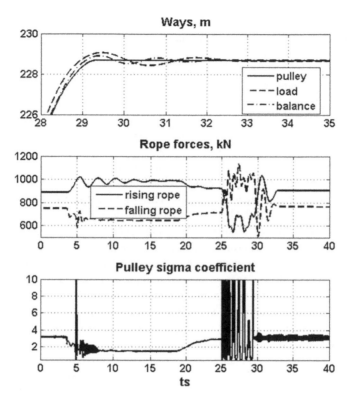

FIGURE 3.34   Safety braking with the elastic rope and rope forces.

$$S_{up} = g\left[(1+k_1)G_l + G_s + p(H+h_1+h_2)\right]$$
$$= 9.81\times\left[1.1\times9.3+8.9+9.11\times(740+2\times22.5)\times10^{-3}\right]= 258 \text{ kN,}$$

(3.48)

For the rope branch without load

$$S_d = g\left[-k_2G_l + G_s + p(H+h_1+h_2)\right]$$
$$= 9.81\times\left[-0.05\times9.3+8.9+9.11\times(740+2\times22.5)\times10^{-3}\right]= 153 \text{ kN}$$

(3.49)

Here the coefficients k_1, k_2 take into account additional losses when skips move, and it is assumed for skip hoists $k = k_1 + k_2 = 0.15$ so that further it is taken $k_1 = 0.1$, $k_2 = 0.05$.

The static force of the hoisting system, which is equal to the difference between the tensions of the loaded and idle rope branches

$$F = g(1+k)G_l = 9.81\times1.15\times9.3 = 105 \text{ kN.}$$

(3.50)

The IM with power of 1400 kW and with $Z_p=6$ is preselected. Its rated torque is $M_{nom}=27.1$ kN-m and the moment of inertia $J_m=173.4$ kg-m. Since the drum maximum rotational speed is $N_d = \dfrac{10 \times 30}{2.5 \times \pi} = 38.2$ rpm, supposing that the motor will operate with a maximum frequency of about 40 Hz, the gear ratio is $i = 500 \times 0.8/38.2 = 10.5$. The gear moment of inertia reduced to the output shaft is $J_i=22.75$ t-m², its efficiency $\eta_i=0.96$.

The total equivalent mass of the hoisting system is $M_{ekv}=9.3+2 \times 8.9+2 \times 9.11 \times 970 \times 10^{-3}+9.11 \times 785 \times 10^{-3}+2 \times 4 \times 18/4.95^2+(140+22.75+173.4 \times 10.5^2 \times 10^{-3})*4/5^2=87$ t. In this expression, the terms, in the order of their places in the formula: load, two empty skips, two branches of the head rope, tail rope, two guide pulleys, drum, gearbox, motor. Thus, taking the total efficiency of all rotating elements equal to 0.92, the motor torque during the steady movement is $105 \times 2.5/(10.5 \times 0.92)=27.2$ kN-m, and during acceleration of 1 m/s² $M_{max}=27.2 \times (1+87/105)= 49.7$ kN-m. Therefore, $M_{max}/M_{nom}=1.83$.

The motor speed diagram when lifting a load consists of five sections. In the first section, the skips accelerate; the unloaded skip leaves the unloading zone, the length of which is h_p. The speed in this section should not exceed 1 m/s. Thus, the duration of the passage of this section is $2 \times 2.17/1=4.34$ s. In the second section, the skips accelerate from a speed of 1 m/s to a speed of 10 m/s in 9 s, covering a distance of 49.5 m. In the fourth section, the skips slowdown from a speed of 10 m/s to a speed of 1 m/s, and if the acceleration and deceleration values are equal, then the skips also cover a distance of 49.5 m. On the fifth section, the loaded skip enters the unloading zone. Then the path that the skip passes at a steady speed is $785-2 \times 2.17-2 \times 49.5=681.7$ m. In fact, to ensure that the speed drops to 1 m/s by the time the skip enters the unloading zone, slowing down can start a little earlier. If the jerk limiting mode is used, then the velocity diagram becomes more complicated, and it is advisable to refine it based on the simulation results, which also make it possible to take into account possible distortions in the formation of the given acceleration schedule caused by the imperfection of the control system. Note that the process of unloading the skip begins on the fifth section, so instead of (3.50), we have

$$F_5 = g\left(1+k-\alpha_p\right)G_l \tag{3.51}$$

where $\alpha_p=0.3$–0.4 is the coefficient that takes into account the degree of skip emptying when it stops in the unloading zone [10].

In the **Mine3a** model, the motor is powered by a three-level inverter. The system of direct vector control is applied, as in the **Paper2a** model, in the description of which the corresponding explanations are given. For the convenience of subsequent consideration, the entire installation is considered three-mass, but with rigid connections: a motor with a gearbox and a drum, including the unused part of the rope and assuming that the string is 60 m long, the ascending rope branch with a load together with the guide pulley, the descending rope branch without a load, also with guide pulley. Based on the above expression for the equivalent mass of the lifting system M_{ekv}, we have $m_1=38.5$ t, $m_2=28.9$ t, $m_3=19.6$ t. This value of m_1 corresponds to the

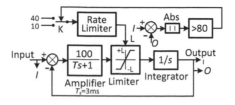

FIGURE 3.35   Diagram of the rate limiter with a limitation of the acceleration slew rate.

equivalent moment of inertia of the motor $J_{meqv} = \dfrac{38,500 \times 25}{4 \times 10.5^2} = 2182$ kg-m^2. Motor

load torque $T_l = \dfrac{D_d}{2i}\left(S_{up} + am_2 - (S_d - am_3)\right)$, where a is an acceleration of the load lifting.

The model uses a rate limiter developed specially, with a limitation of the acceleration slew rate, the diagram of which is shown in Figure 3.35. The rate of change of the output signal is determined by the limit value L. At the beginning of speed command fabrication, the difference between input I and output O is large, the acceleration value corresponds to 1 m/s^2 (40 rpm/s), but this value is not set immediately, but with a rate that is determined by the internal rate limiter, for 2 s in the case. When the process of changing the speed reference is close to the end, the value of O approaches the value of I, the acceleration value gradually decreases.

In the model, the motor starts to speed up with a small acceleration to a speed corresponding to 1 m/s until the skips leave the loading/unloading areas. Then the motor speeds up to a speed corresponding to 10 m/s with increasing acceleration as described above. The moments of deceleration (more precisely, the heights of the rise when these moments occur) are determined by the constants D1− the beginning of deceleration to a speed of 1 m/s, D2− the beginning of deceleration to a stop, D3- the application of the brake. The choice of small values of these quantities makes it possible to adjust the required speed diagram quickly. For example, if to set D1 = 110 m, D2 = 175 m, D3 = 177 m, the process shown in Figure 3.36 is received. It can be seen that at $t=5$ s, with a speed of 1 m/s, the skip leaves the loading zone and starts accelerating, reaching a lifting height of 110 m at $t=20$ s. After that, the load starts to slow down, and at $t=34$ s, it reaches a speed of 1 m/s. At this speed, it remains for 3.5 s, after which it enters the unloading zone and stops, at this having traveled a distance of 2 m, the total height of the lift is 177 m. It can also be seen from the plots that the specified acceleration change diagram is fulfilled, providing jerk limitation.

Since in the considered model the lifting height is 785 m, then, to obtain the corresponding results, it is necessary to add 785–177 = 608 m to the values D1, D2, D3, that is, set D1 = 718 m, D2 = 783 m, D3 = 785 m The resulting process is shown in Figure 3.37.

Similar results can be obtained when using a simple rate limiter with a first-order lag element at the output to limit the jerk, the toggle switch in the model is in the down position. In this case, however, it turned out to be necessary to slightly correct the values D1 = 100 m and D2 = 174.5 m to obtain a lift height of 177 m. The process

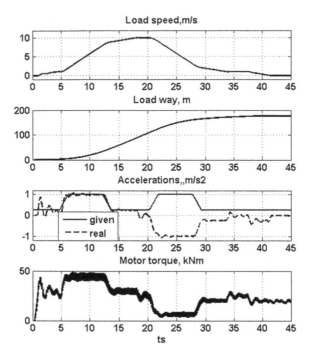

FIGURE 3.36   Skip lifting with decreased heights.

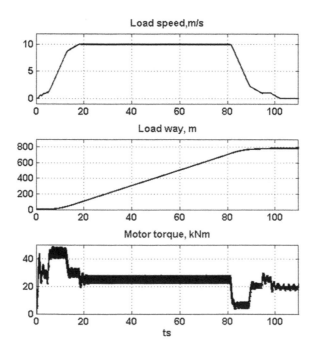

FIGURE 3.37   Skip lifting at the complete height.

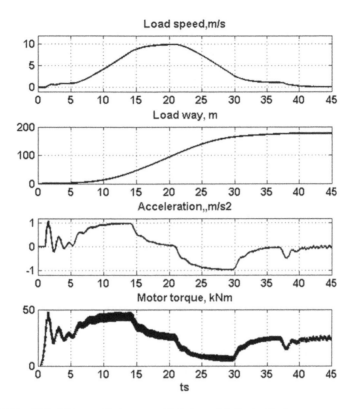

FIGURE 3.38 Skip lifting with decreased heights, with limited jerk.

is shown in Figure 3.38. It can be seen that the acceleration changes smoothly, and the value of the jerk is limited.

The active rectifier fabricates voltages 2×2500 V in DC link in this model; the rectifier is powered by the network 35 kV through a transformer 35/2.8 kV. Changes of the network powers and the network current shape when the load is lifted to the height of 177 m are shown in Figure 3.39. The reactive power is close to zero.

The following models take into account the elasticity of the head ropes. First, we remain within the framework of the three-mass model. In this case, various options for taking into account the mass of the guide pulleys and the residual length of the ropes are possible. Let us add the mass of pulleys together with a string of ropes, which has a relatively short length and is considered inextensible, to the motor with a drum and a gearbox. Then, keeping in mind that when lifting the load, the total mass of the rope on the drum remains constant, and taking into account the presence of a tail rope, it will be $m_1 = 9.11 \times (2 \times 970 - 785) \times 10^{-3} + 2 \times 4 \times 18/4.95^2 + (140 + 22.75 + 173.4 \times 10.5^2 \times 10^{-3}) \times 4/5^2 = 45.5$ t, $m_2 = 9.3 + 8.9 + 9.11 \times 10^{-3} \times 785 = 25.4$ t, $m_3 = 16.1$ t.

In this case, the segments of the ropes on the drum are considered inelastic, since their elongation is prevented by friction on the surface of the drum. As it can be seen, in the adopted model, there is a deviation from the previously applied principle,

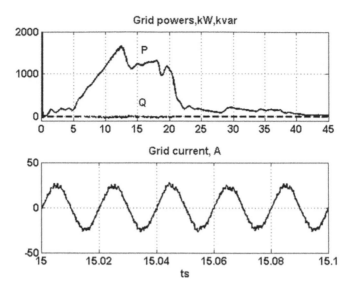

FIGURE 3.39 Network powers and current when the loaded skip is lifted.

according to which the rope elasticity is taken into account by increasing the point mass by 1/3 of the mass of the rope, which is connected with the mass. This circumstance is caused by the fact that the mass m_1 is modeled by an induction motor with the corresponding moment of inertia, in the model of which the latter does not change during operation and requires a significant complication of the model. It can be assumed that the results will not differ qualitatively and to a large extent quantitatively. Note that the specified value m_1 corresponds to the moment of inertia of the motor $J_{meqv} = 2573$ kg-m².

In next models, the cross-sectional area of the rope is 911 mm², so that the linear stiffness of the rope is $K = 1.2 \times 105 \times 911 = 1.1 \times 10^8$. In the **mine3b** model, the same process is modeled as in the **mine3a** model (Figure 3.36), the resulting plots are shown in Figure 3.40. It can be seen that at the beginning of the lifting, the oscillations of the elastic torque of the descending, idle rope branch have a lower amplitude and a higher frequency, since the length of the branch is much less. Consideration of the plot of the load position on a large scale shows that it stops with an error of about 0.1 m, which is considered acceptable.

In the following models, the simulated lifting system is considered a 5-mass one: a motor with a drum and a gearbox, two skips, and their guide pulleys. For convenience, these masses are indicated by numbers, as shown in Figure 3.41. Assuming the tail rope is not elastic and the rope length between the pulley and drum is 60 m, the equivalent mass values are equal to:

$$m_5 = \left(140 + 22.7 + 173.4 \times 10.5^2 10^{-3}\right) \times \frac{4}{5^2} + 9.11$$

$$\times \left(2 \times (970 - 60 + 20) - 785\right) \times 10^{-3} = 38.9 t$$

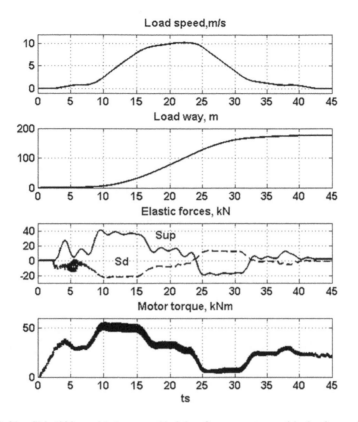

FIGURE 3.40   Skip lifting with decreased heights, 3-mass system, with elastic consideration.

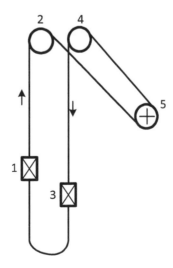

FIGURE 3.41   5-mass hoisting system.

$$m_1 = 9.3 + 8.9 + 9.11 \times 10^{-3} \times \left(x + \frac{1}{3} \times (785 - x) \right) = 20.6 + 6.1 \times 10^{-3} x t$$

$$m_2 = \frac{18 \times 4}{4.95^2} + 9.11 \times 10^{-3} \times \left(20 + \frac{1}{3} \times (785 - x) \right) = 5.5 - 3 \times 10^{-3} x t$$

$$m_3 = 8.9 + 9.11 \times 10^{-3} \times \left(\frac{1}{3} x + (785 - x) \right) = 16.1 - 6.1 \times 10^{-3} x t$$

$$m_4 = \frac{18 \times 4}{4.95^2} + 9.11 \times 10^{-3} \times \left(20 + \frac{1}{3} x \right) = 3.1 + 3 \times 10^{-3} x t$$

The behavior of the system is described by the following equations. The motor:

$$J_{eq} \frac{d\omega_m}{dt} = T_m - \frac{D_d}{2i} (F + F_{25} - F_{45}) \tag{3.52}$$

Here J_{eq} is the equivalent moment of inertia equal to $J_{eq} = \frac{D_d^2 m_5}{4i^2} = \frac{25 \times 38.9 \times 10^3}{4 \times 10.5^2} =$ 2205 kg-m², ω_m, T_m is the rotational speed and the torque of the motor, $F = gG_l = 91.2$ kN, F_{25}, F_{45} are elastic forces from the guide pulleys caused by the deformation of the ropes.

Guide pulleys:

$$m_2 \frac{dV_2}{dt} = F_{25} - F_{12} - P_2 \tag{3.53}$$

$$m_4 \frac{dV_4}{dt} = F_{34} - F_{45} - P_2 \tag{3.54}$$

The quantity P_2 takes into account losses under rotation of the guide pulleys and is taken equal to 1 kN.

The skips:

$$m_1 \frac{dV_1}{dt} = F_{12} - P_1 \tag{3.55}$$

$$m_3 \frac{dV_3}{dt} = -F_{34} - P_3 \tag{3.56}$$

The resistance to movement of the skips is assumed to be $P_1 = 0.1G_l = 9300$ N, $P_3 = 0.05 \, G_l = 4650$ N. Elastic forces are determined by formulas similar to 3.43–3.46, with $V_5 = \omega_m \times D_d / 2i$.

Such a system is simulated in the **mine4a** model. The process of the load lifting at the height of 785 m is shown in Figure 3.42. It is seen that not long before reaching the wanted position, the speed is decreased to 1 m/s, and afterward the skip stops exactly in this position. One can see also elastic fluctuations in rope tensions, but

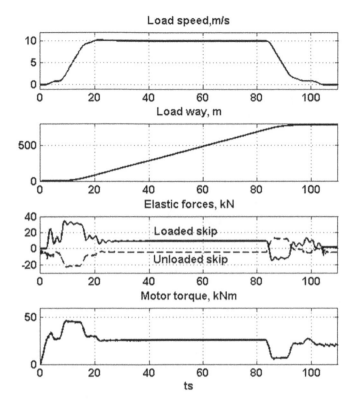

FIGURE 3.42 Load lifting in 5-mass hoisting system.

these fluctuations do not affect the movement of the skips due to their significant masses.

Figure 3.43 shows the changes in active and reactive power and the shape of the network current.

The **mine4b** model simulates the process of the safety braking. At $t = 30$ s, when the loaded skip has risen to a height of 180 m, a braking signal is given. The motor is immediately switched off and a signal is given to apply the brake, its braking force, reduced to the circumference of the drum, is 200 kN, and the dynamic characteristics are indicated earlier. The process of the stop is shown in Figure 3.44. It is seen that slowing down duration is 3 s, therefore, the deceleration is 3.3 m/s². In the process of deceleration and immediately after stopping, fluctuations occur in the forces of the head ropes of significant amplitude, which practically do not affect the process of stopping the skips, but which must be taken into account when choosing ropes.

3.3 BELT CONVEYORS

Belt conveyors are the most efficient means for transporting masses of bulk materials. There is a wide variety of types, configurations, and modifications of belt conveyors depending on the type of transported material, on the installation site (underground

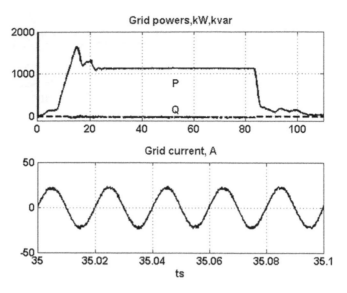

FIGURE 3.43   Powers and the network current under load lifting in 5-mass hoisting system.

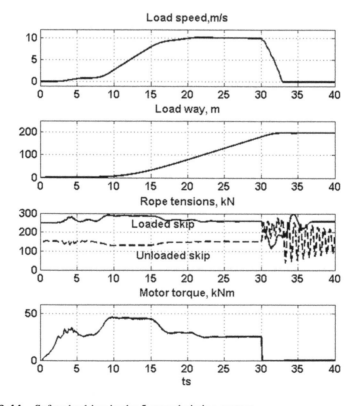

FIGURE 3.44   Safety braking in the 5-mass hoisting system.

or on the surface of the earth), on the terrain, on the required distance of load transported, on the required capacity, and so on. Figure 3.45a shows the layout of a simple
conveyor. The load-bearing and traction element is an endless flexible belt, whose
upper (load-carrying) and lower (idle) branches are carried by roller bearings (belt
idlers), and which envelopes the drive and tail pulleys at the ends of the conveyor.
The movement is transmitted to the belt by a friction from the drive pulley. The necessary initial tension on the idle branch is created by the tension drum with the help
of the tensioner. There are designs in which tension is created by moving the tail pulley. For horizontal conveyors and for conveyors that move the load upward, the drive
pulley is located at the discharge point. If the conveyor is designed to move the load
downward, then the drive pulley is located at the loading point.

Longer conveyors often have more complex layouts. Figure 3.45b shows the drive
end of a conveyor with two drive pulleys placed close to each other. Owing to the
deflecting pulley, an increase of the angle of wrapping the pulley with the belt, as
well as a wrap of the second pulley with the lower (dry, clean) side of the upper belt,
is ensured.

In long conveyors, drive pulleys can be placed along the conveyor [11,12]. The
layout shown in Figure 3.45c is a simplified version described in [12].

Previously, IMs with a squirrel-cage rotor were used for conveyor drives; however,
by the conveyor start, there was a significant increase of the tension, which led to a
rapid wear of the belt, which is a very expensive part of the conveyor. Therefore, IMs
with a phase rotor began to be employed, and a soft start was provided by setting a
chain of resistors in the rotor, and then using the Sherbius cascade. However, recently,
the use of IMs with a squirrel-cage rotor has become widespread again, but which are
controlled by voltage source inverters (VSI). This solves not only the problem of soft
start but also the possibility of saving energy by reducing the speed of the conveyor

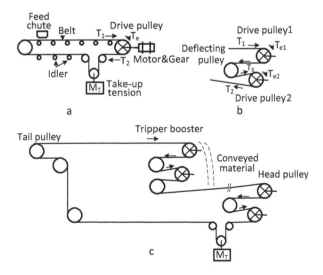

FIGURE 3.45 Possible layouts of conveyors (a) simple conveyor; (b) conveyor with two
drive pulleys; (c) conveyor with the drive pulleys along the conveyor.

during those periods when its nominal capacity is not required. The choice of electric drive power is associated with the calculation of the belt tension, which is a rather difficult task. Modern approaches to such calculations are given in [13,14].

The conveyor shown in Figure 3.45 is considered. Here T_1 is a tension of the load branch in front of the drive pulley, T_2 is a tension of the return branch behind the drive pulley, T_e is a tension which is produced by the motor and which is used to overcome all resistances when the belt moves, $T_e = T_1 - T_2$. It can be written based on (3.34)

$$C_w = \frac{T_2}{T_{emax}} = \frac{1}{e^{f\theta} - 1} \tag{3.57}$$

where C_w is a wrap factor. It depends on the wrap angle θ and on conditions on the pulley and belt surfaces. It is advised to use $C_w = 0.5$ for the wrap angle of 180°. Hence, having the calculated value T_{emax}, the minimal value of the tension T_2 may be found. It should be noted that this tension should ensure that the belt sags between the support rollers are not more than the specified values; the relevant formulas are given in the referenced literature. Thus, in the case under consideration, the maximum tension of the belt is 1.5 T_e.

Reducing this tension can be reached with the increase of the wrap angle by employment of the deflecting pulley. If, for example, the wrap angle is increased to 210°, it is advised to take $C_w = 0.38$, and the belt maximum tension is 1.38 T_e.

An essential decrease of the equivalent C_w value is achieved by using of two drive pulley (Figure 3.45b). Designate $T_e = T_{e1} + T_{e2}$, $T_{w1} = T_3/T_{e1}$, $C_{w2} = T_2/T_{e2}$. Eliminating from these equalities $T_3 = T_2 + T_{e2}$, we find

$$C_w = \frac{T_2}{T_e} = \frac{C_{w1}C_{w2}}{1 + C_{w1} + C_{w2}} \tag{3.58}$$

Let, for example, the wrap angle of the first pulley is 180° ($C_{w1} = 0.5$), and the second 210° ($C_{w2} = 0.38$) (Figure 3.45b). Then the equivalent value of C_w is 0.1, that is, the belt maximum tension is 1.1 T_e. A power ratio of the first and second pulley motors are

$$\frac{P_1}{P_2} = \frac{T_{e1}}{T_{e2}} = \frac{1 + C_{w2}}{C_{w1}} \tag{3.59}$$

When several drive pulleys are placed along a conveyor line, the problem arises of dividing the load between them. This task is easier to solve for pulleys on the lower (unloaded) branch, since here the operating conditions are almost constant. On the upper branch, there may be moments when the loading of individual sections of the conveyor can be completely different [11]. Consider a hypothetical example. Take the total value of T_e as 100%. Let the tension at the tail pulley be 20%. Then the tension changes along the upper branch, when the conveyor is loaded completely, which can be depicted as line 1 in Figure 3.46. The pulley motor power in pu is equal to tension value in pu, since the belt speed in all points are supposed to be the same, P_H is the demanded power of the head pulley motor in pu. In the middle of the conveyor

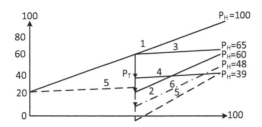

FIGURE 3.46 Tension and power change along the conveyor under the different load conditions.

belt, the drive pulley is installed, whose motor produces 40% of the total power, P_T is its power. Then the belt tension will change as it is shown by line 2.

Suppose now that at some point in time, the first half of the upper belt branch is fully loaded, and the second is empty [11]. Let us assume that the linear weight of the belt is equal to 15% of the nominal distributed weight of the transported material, that is, 13% of the total weight loading the conveyor. If the intermediate pulley motor does not turn on, then the tension will change as shown by line 3. If this motor is turned on, and the power distribution of the motors is the same as with the nominal load of the conveyor (60/40), then the change of tension is shown by line 4.

Now the second half of the upper branch is fully loaded, and the first is empty. Since the total required power is still equal to 65%, if to maintain the system settings, the tension after booster will drop to very small, theoretically, to negative values [11] (dashed dependence 5), which leads to instability of the process, slipping of the belt along the surface of the pulley and increases the intensity of destruction of the supporting pulleys and belt. It is obvious that such phenomena must be avoided. To provide the conditions for the required tension of the belt along the entire length, it is proposed to turn off the booster when the conveyor is loaded much less than the maximum and turn it on when the total required power exceeds a certain value b_0. If to designate P_0—the required total power, g_h, g_t—load distribution coefficients, $g_h + g_t = 100$, then the relative powers of the motors are determined by the relations:

Under $P_0 \leq b_0$, $P_H = P_0$, $P_T = 0$.

Under $P_0 > b_0$,

$$P_H = \frac{g_h(P_0 - b_0) + b_0(100 - P_0)}{100 - b_0}, \quad P_T = \frac{g_T(P_0 - b_0)}{100 - b_0}. \tag{3.60}$$

For $b_0 = 40\%$, plots of the power changes are given in Figure 3.47. For $P_0 = 65\%$, $P_H = 48.3\%$, $P_T = 16.7\%$. The tension changes as it is shown by dash-and-dot line 6 in Figure 3.46.

Firstly, a simple horizontal conveyor, as in Figure 3.45a is considered. Its parameters: the length $L = 800$ m, the capacity $Q = 3400$ t/h, the speed $V = 3$ m/s, the belt width 1200 mm, the belt specific mass $m_b = 23$ kg/m.

In order to select a motor, it is necessary to calculate the tension T_e that this motor must create to move the conveyor with a given speed and load. This tension has a number of components, the calculation of which is described in [13,14].

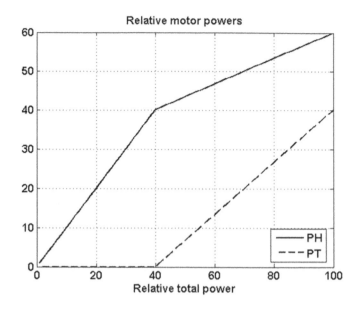

FIGURE 3.47 Power changes when the booster is turned off at the small conveyer load.

Here, these calculations are somewhat simplified. The main component is the resistance caused by the movement of the belt along the conveyor elements, T_H and T_D are its components for the upper and lower branches of the belt, respectively, $T_{el} = T_H + T_D$.

$$T_H = Lgf\left(m_{RH} + m_b + m_G\right) \tag{3.61}$$

$$T_D = Lgf\left(m_{RD} + m_b\right) \tag{3.62}$$

$$m_G = \frac{Q}{3.6V} \tag{3.63}$$

f is a friction coefficient that depends on a number of factors and usually is taken in the range of 0.02–0.03.

m_{RH} and m_{RD} are liner loads of the rotating elements of the roll supports, for upper and lower belt branches, respectively:

$$m_{RH} = \frac{G_H}{l_H}, m_{RD} = \frac{G_D}{l_D} \tag{3.64}$$

where G_H and G_D are masses of the rotating elements of the roll supports, l_H and l_D are the distances between roller on the upper and lower branches, respectively. Values of G_H and G_D are determined by the conveyor construction. Using data given in [15], when the belt width is 1200 mm, one can take $G_H = G_D = 30$ kg. We take $l_H = 1$ m, $l_D = 3.3$ m. It gives $m_G = \dfrac{3400}{3.6 \times 3} = 309$ kg/m, $m_{RH} = 30$ kg/m, $m_{RD} = 9.1$ kg/m,

$T_H = 800 \times 9.8 \times 0.025 \times (30 + 23 + 309) = 70.96 \, kN$, $T_D = 800 \times 9.8 \times 0.025 \times (9.1 + 23) = 6.29 \, kN$, $T_{el} = 77.25$ kN.

The forces T_p must be added to this value which is required to rotate the non-drive pulleys. In accordance with the data given in [13], for pulleys with a wrap angle of more than 150°, the force for pulley rotation is estimated at 750 N, and with a smaller angle, at 500 N. Then, in accordance with Figure 3.45, $T_p = 2 \times 750 + 2 \times 500 = 2.5$ kN.

When the material falls on the belt, the component of this velocity in the direction of the belt movement V_0 is close to zero, and after hitting the belt, it acquires a velocity V, for which it is necessary to apply a force T_{am}, the value of which can be found from the relation

$$T_{am} = \frac{Q(V - V_0)}{3.6} \qquad (3.65)$$

Taking $V_0 = 0$, we find $T_{am} = 3400 \times 3/3.6 = 2.8$ kN. Thus, in static $T_e = 82.6$ kN. The maximum tension $T_1 = T_e + T_2$. The minimal value of T_2 should provide an acceptable value of belt sagging of no more than 3% [13], from which

$$T_{20} = 4.2g(m_b + m_G)l_H = 4.2 \times 9.8 \times 332 \times 1 = 13.7 \text{ kN}. \qquad (3.66)$$

The value of the tension T_2 must ensure that the required torque is transmitted by the drive pulley. Taking the wrap angle equal to 180°, by (3.57) $T_2 = 41.3$ kN, $T_1 = 82.6 + 41.3 = 123.9$ kN.

Assuming, in accordance with [13], the force to overcome losses in the drive pulley is 1 kN and the efficiency of the mechanism is 0.95, we find the motor power as $P = (82.6 + 1) \times 3/0.95 = 264$ kW.

It should be noted that the motor selected power is significantly affected by the possible presence of even small rise in the direction of transporting the material, since when lifting to a height H, the required tension increases by $gm_G H$. Let, for example, with a conveyor length of 800 m, the rise height is 30 m, those the angle of elevation is only 2°. In this case, the T_e value will increase by $309 \times 9.8 \times 30 = 91$ kN, that is, it will be more than double. In consideration of this fact, the motor is selected with a margin.

Taking the diameter of the drive pulley equal to 1 m, its rotational speed is 6 rad/s ≈ 60 rpm. IM with power of 355 kW and voltage of 400 V, $Z_p = 4$ is chosen. IM has parameters [1]: $M_{nom} = 4562$ N-m, $N_{nom} = 743$ rpm, cos $\varphi = 0.83$, $I_{nom} = 641$ A, the starting current 6.8 pu, the maximum torque 2.5 pu, the moment of inertia $J_m = 21$ kg-mm². IM rotates the pulley with the gearbox with $i = 11.5$, providing the nominal tension $4562 \times 11.5/0.5 = 105$ kN with the rotational speed of 660 rpm.

A calculation of the conveyor equivalent mass is carried out. The IM moment inertia reduced to the pulley shaft is $J_m^* = 21 \times 11.5^2 = 2777$ kg-m². The gearbox moment of inertia reduced to the slow axis is taken as $0.2 J_m^* = 555$ kg-m². By data in [13], the mass of the pulley, with a diameter of 1 m and belt width of 1200 mm, may be estimated as 600 kg, its moment of inertia as $J_p = 150$ kg-mm². This applies to the head and tail pulleys. Diameters of the other three pulleys are taken equal to 0.75 m, they have the mass of 400 kg and a moment of inertia 56 kg-m. Therefore, the equivalent mass of the drive system is $m_r = (2777 + 555 + 150)/0.25 = 13{,}928$ kg ≈ 14 t, the equivalent mass of the loaded conveyor $600 + 3 \times 400 + 800 \times (30 + 223 + 309 + 9.1) = 317$ t, and non-loaded one $600 + 3 \times 400 + 800 \times (30 + 223 + 9.1) = 69.8$ t.

Start occurs usually with the unloaded conveyor; however, it can be situations, when the loaded conveyor has been stopped, and afterward, restart is demanded. Therefore, the start of the loaded conveyor is under consideration. It is considered that the belt tension at start-up should not exceed 170% of the rated static tension, that is, in this case $1.7 \times 124 = 210$ kN. As shown above, with a drive pulley wrap angle of 180°, the tension created by the motor should not exceed $210/1.5 = 140$ kN, that is, the dynamic tension should not exceed $140 - 83 = 57$ kN, so that the acceleration should not exceed $57/(317 + 14) = 0.172$ m/s², the acceleration duration should not be less than $3/0.172 = 17.4$ s. The value of T_2 must be 70 kN. The IM torque during speeding up is $(83.6 + 57) \times 0.5/0.95/11.5 = 6.44$ kN-m which the selected IM can provide.

This system is simulated in the **Conv1** model. The DTC **AC4** system is used to control IM. VSI supply is carried out with a diode rectifier so that the operation mode with energy recovery is not possible. Since various forms of rotational speed change are investigated during the simulation, the blocks for their implementation (rate limiter and inertial element) are installed outside **AC4**; a high rate of output change is set in the rate limiter that is present in **AC4** which does not affect the rotational speed change practically. In the **Conveyor** subsystem, tension changes are modeled separately for the upper (loaded) and lower (idle) branches of the belt, taking into account the tensions that occur when the conveyor speed changes, with separate consideration of the branch equivalent masses, the values of which are given above. In this case, the tail pulley is assigned to the upper branch, and the rest to the lower one. The equivalent moment of inertia of the motor includes the moments of inertia of the motor, gearbox, and head pulley. Scope **Scope_tens** fixes also the value of the ratio T_e/T_2, which defines slippage absence on the head pulley and which, with the selected parameters, should be less than 2.

Changes of the motor torque and tensions during acceleration, steady motion and braking of a conveyor with a load are shown in Figure 3.48. The acceleration and deceleration values are set to 0.15 m/s². For comparison, the change in tension T_1 in the absence of a lag element at the output of the input rate limiter is shown in Figure too. It can be seen that in this case, the tension at start-up increases faster and with some overshoot. The condition of no slip on the head pulley is satisfied.

Figure 3.49 shows changes of the torque and tensions during the start and slowing down of the unloaded conveyor that is a more usual procedure.

It is interesting to find out how the ability to control the speed affects energy savings. The simulation shows that with a full load at a steady speed of 3 m/s, the power consumed from the grid P_g is 270 kW. If the capacity is halved, then the value of m_G will also be halved. At this, according to the simulation results, $P_g = 170$ kW. If, however, when Q is halved, the speed is also halved, then the value of m_G will not change. If the simulation is carried out at a conveyor speed of 1.5 m/s, then the value of $P_g = 140$ kW, that is, the effect of saving energy is obvious.

In fact, the conveyor belt has noticeable elastic properties, and its movement, especially during start-up, has a wave character [16]. Exact modeling of such a

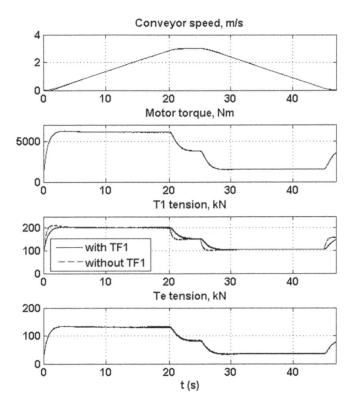

FIGURE 3.48 Motor torque and tensions during acceleration, steady motion and braking of a simple conveyor with a load.

system is difficult; therefore, the finite element method is used to take into account the elastic properties of the belt [17]. Only longitudinal elastic reactions are taken into consideration.

In this method, the belt is divided into $2n$ equal segments (n for the upper and n for the lower branches), each segment is considered an inelastic body, and the segments are connected to each other by a weightless spring with an elasticity coefficient C and a damping coefficient $d = k_d C$. The same approach has been applied previously in the simulation of mine hoists but with a much smaller number of individual masses.

Figure 3.50 shows a conveyor model. It is supposed that the tensioner is located near the head pulley, and that together they form a single rigid system. The numbering of segments starts from this system toward the tail and then toward the head pulley. The movement of each segment is described by the equation

$$m_{H(k)} \frac{dV_i}{dt} = T_i - T_{i-1} - T_{lH(k)} \tag{3.67}$$

Here m_H or m_k are masses of one segment of the upper or lower belt branches, respectively, which are equal to the total branch mass with load divided by n, T_i is an elastic force which acts on this segment by the following segment and assists to move,

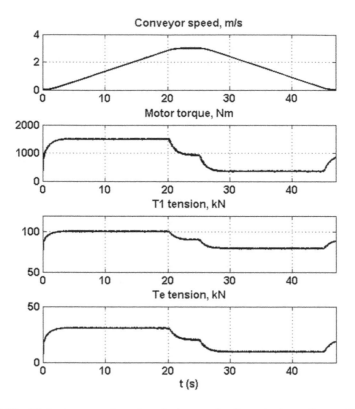

FIGURE 3.49 Motor torque and tensions during acceleration, steady motion, and braking of a simple conveyor without a load.

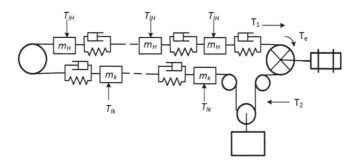

FIGURE 3.50 Conveyor model for investigations of the elastic properties.

T_{i-1} is an elastic force which acts on this segment by the preceding segment and impedes to move, T_{lH} or T_{lk} are forces of resistance to the movement of the segment in the upper or lower branches, respectively, equal to the total resistance to movement of the branch, divided by n. Then

$$\frac{dx_i}{dt} = V_i \tag{3.68}$$

$$T_i = C(x_{i+1} - x_i) + d(V_{i+1} - V_i) \qquad (3.69)$$

For the first segment $T_{i-1} = T_2$, for the last one (nearest to the motor) $T_i = T_1$, V_{i+1} is the circumferential speed of the head pulley.

In the **Conv1b** model, the motor is simulated by a simplified way, as in the **Crane4** model, which allows estimating quickly the influence of the law of speed change on the processes in the conveyor; this procedure requires a lot of simulation time; complete model is **Conv1c**. The number of elements in the belt branch $n = 8$, and if desired, can be easily increased by inserting additional segments of the same structure, the number n is specified in the *Model Properties/Callbacks/InitFcn* field. When starting the conveyor, an initial tension must be created by turning the head pulley at a certain angle. In the model, the initial tension is created by increasing the value of x_{i+1} relative to x_i by dX the value of which is specified in the same field.

The model provides for the possibility of studying the influence of the speed setting option on the tension change at start-up. With the value of the constant **Var_speed** = 1, the same circuits are implemented as in the previous model: a rate limiter with constant acceleration plus a first-order inertial element. With **Var_speed** = 2, the acceleration a changes according to the law

$$a = \frac{\pi}{2} \frac{V_{ref}}{T} \sin \frac{\pi t}{T} \qquad (3.70)$$

where T is the accepted speeding-up time of the conveyor [18]. When **Var_speed** = 3, a complicated rate limiter with variable acceleration is implemented (Figure 3.35).

It is taken for simulation $C = 6.5 \times 10^6 / L$ N/m, $k_d = 0.4$. Next, several processes are simulated. Changes of the tension T_1 and ratio $T_e / T_2 = C_w^{-1}$ for two values of the speeding up time: $T = 20$ s and $T = 40$ s, are shown in Figure 3.51. It is seen that in the first case, the condition of no slippage on the main pulley is not met, while with an increased speeding up time, this condition is met. In this case, a decrease in the maximum tension value is also observed. The same process for the third variant of the speed setting is shown in Figure 3.52, when the speeding up time is 40 s and the time of the acceleration rise is about 20 s. It can be seen that it almost does not differ from that shown in the previous figure with the same speeding up time. When running the simulation for the second variant for setting the speed, one can see that the results are somewhat worse. Thus, it is reasonable to use the first variant as it is easier to implement.

The start of the unloaded conveyor is shown in Figure 3.53, the speeding up time is 20 s. It can be seen that it goes on quite smoothly. Thus, it may be reasonable when starting an unloaded conveyor, which is a common procedure, to apply faster ramp times than when starting a loaded conveyor, which usually occurs after an unexpected stop.

The process of acceleration and deceleration for the complete model **Conv1c** at $T = 40$ s is shown in Figure 3.54. It can be seen from a comparison with Figure 3.51 that the plots of tension and ratio T_e / T_2 are almost identical, although the simulation in the model under consideration is much slower.

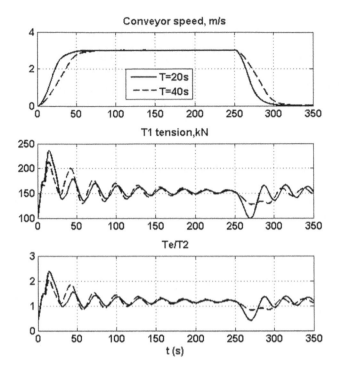

FIGURE 3.51   Conveyor operation with the first variant of the speed setting.

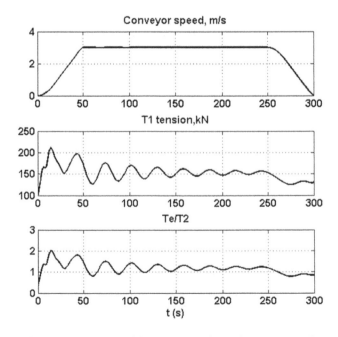

FIGURE 3.52   Conveyor operation with the third variant of the speed setting.

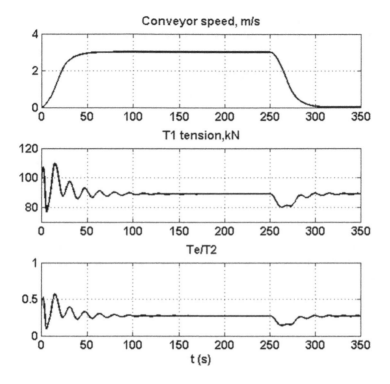

FIGURE 3.53   Unloaded conveyor operation.

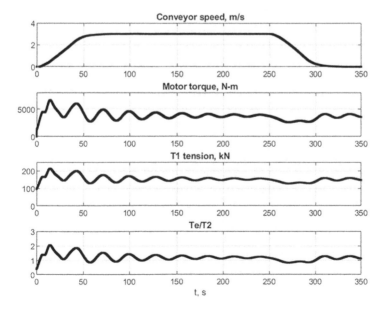

FIGURE 3.54   Conveyor operation with the first variant of the speed setting, complete model.

The following models simulate a conveyor with two drive pulleys, as in Figure 3.45b. In order not to repeat the calculations, we will assume the same parameters of the conveyor as in the previous models, with the difference that this is an inclined conveyor with material fed upward to a height of 30 m. As already mentioned, the increase in the required tension will be 91 kN, that is, $T_e = 83.6 + 91 = 174.6$ kN, and the required static power of the motors is $174.6 \times 3/0.95 = 551$ kW. According to (3.59), assuming the wrap angles indicated above, ratio of motor power $P_1/P_2 = 1.38/0.5 = 2.76$. For the purposes of unification, for the drive of the first pulley, two identical IM is used, and for the second pulley, one such IM is used. IMs with the rated voltage of 400 V and 4 pole pairs are used which drive pulleys with the gearboxes with $i = 11.5$. IM parameters: power 250 kW, speed 743 rpm, efficiency 94.5%, cos $\varphi = 0,80$, rated current $I_{nom} = 472$ A, starting current 7.5 pu, rated torque $M_{nom} = 3213$ N-m, starting torque 1.6 pu, maximum torque 2.7 pu, $J_m = 13,5$ kg-m². Taking into account the dynamic force of 57 kN calculated earlier, we have $T_e = 232$ kN; it was found earlier that the resulting coefficient $C_w = 0.1$; therefore, the minimal value of $T_2 = 24$ kN, which is more that value T_2 for acceptable belt sag calculated earlier as 13.7 kN, formula (3.66). So, the minimum T_1 value is 256 kN, we take $T_2 = 30$ kN.

When creating a model of this conveyor, two tasks need to be solved. For the first pulley, two identical motors are used, connected in parallel and powered by one inverter, while the models under consideration use the **AC4** library model with one motor. It is obvious that if the motors and their kinematic connections are identical, they can be replaced by one motor of double power by modeling the conveyor, the parameters of which in pu are the same as the parameters of one IM. This approach is applied in the developed models. If it is of interest to simulate motors that have slightly different characteristics or different kinematic connections, it will be necessary to create a special model, which is not considered here.

Next, it is necessary to simulate the distribution of load torques between the motors in accordance with the reference. This is achieved in the same way that was used in the **Paper2e** model (in a different circuit design). Since the circumferential speeds of both drive pulleys can differ slightly, in the model, with a small difference in these speeds, the load torques are redistributed, at which the motor with a higher speed increases its load, and the motor with a smaller speed decreases the load, so that the sum of the loads does not change.

The process of starting a conveyor is simulated in the **Conv2** model; the rate limiter with a constant acceleration of 0.15 m/s² plus a first-order lag element with a time constant of 5 s is used for the fabrication of the speed reference. Both **AC4** systems are powered from the same 400 V source through a 400/480 V transformer. The first system operates in speed control mode, and the second (second pulley) in motor torque control mode, as provided by the **AC4** circuit diagram. The output of the speed controller of the first motor sets the torque of this motor T_{m1} and, with the coefficient k_{sh}, the torque of the second motor T_{m2}. If to designate as T_l the total load torque, then $T_l = T_{m1} + T_{m2} = T_{m1}(1 + k_{sh})$, and with $k_{sh} = 0.55$ $T_{m1} = 0.645T_l$, $T_{m2} = 0.355T_l$. In the model, the value k_{sh} increases from 0.55 to 0.66 at $t = 40$ s.

In the **Conveyor** subsystem, the tensions of the upper T_H and lower T_D branches are calculated separately, as in the **Conv1** model, taking into account dynamic components, and taking into account the increase in tension T_H for lifting the load.

The tension between pulleys T_3 is computed in the model by two ways: $T_{3-1} = T_1 - T_{e1}$ and $T_{3-2} = T_2 + T_{e2}$ where T_{e1}, T_{e2} are forces produced with the motor torques. The scope **Scope_Cw** records values $C_{w1}^{-1} = \dfrac{T_1 - T_3}{T_3}$, $C_{w2}^{-1} = \dfrac{T_3 - T_2}{T_2}$, which must be less as $1/0.5 = 2$ and $1/0.38 = 2.63$, respectively, and also the result value $T_H C_w^{-1} = \dfrac{T_1 - T_2}{T_2}$.

Simulation with $T_2 = 30$ kN showed that by the acceleration, the coefficient C_{w2} reaches a critical value of 0.37, that is, slippage of the belt on the second pulley is possible, especially when the weather conditions worsen. Therefore, the simulation was repeated with $T_2 = 40$ kN. The start process is shown in Figure 3.55. It is seen that the torque of the first motor at $t = 20$ s is 6650 N-m, at $t = 41$ s 4800 N-m, and of the second motor 3660 and 3220 N-m, respectively, therefore, the share of the first motor is $6.65/(6.65 + 3.66) = 0.645$ at $t = 20$ s and $4.8/(4.8 + 3.22) = 0.6$ at $t = 41$ s that exactly corresponds to the calculated values; thus, the applied load distribution modeling scheme works quite satisfactorily. The tension at start-up changes quite smoothly, without noticeable jerks. The coefficients C_{w1} and C_{w2} are within acceptable limits with a certain margin. For example, at $t = 20$ s $C_{w1} = 0.87$, $C_{w2} = 0.5$, so it must be by

$$(3.58) \quad C_w = \frac{0.5 \times 0.87}{1 + 0.5 + 0.87} = 0.18;$$ it can be seen on the third channel of the oscilloscope **Scope_Cw** (recall, the reciprocal values are fixed there).

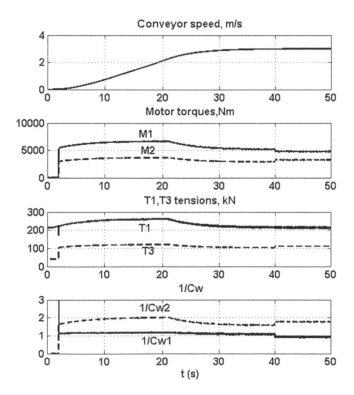

FIGURE 3.55 Start process of the conveyor with two drive pulley, with unelastic belt.

Conveyor belt elasticity is taken into account in the **Conv2a** model. The start process also is simulated in this model, but the time of speeding up is increased to 40s, and the time constant of the lag element to 10s, as in the **Conv1c** model, from which the model of the elastic belt is taken. Start when $n=8$ is shown in Figure 3.56. Damped vibrations of the motor torques and tensions can be seen, the motor torque distribution is up to assignment. Maximum value of the tension $T_1=310$ kN, whereas it is only 250 kN, without elastic consideration. The quantity C_{w2} is approaching to the critical value: $C_{w2max}=0.4$.

It is of interest to investigate the influence of the accepted number of belt segments on the dynamic characteristics. In the **Conv2b** model, the number of segments of one belt branch is $n=16$. The starting process is shown in Figure 3.57. One can see some increase in tension $T_1=324$ kN compared to the previous model, as well as a decrease of C_{w2} to 0.385.

In conclusion, modeling a conveyor with a booster device, as in Figure 3.45c is considered. The parameters of the conveyor are the same as in previous models, with the difference that its length is 2000m, the booster is installed in the middle of the upper branch. The tension of the lower branch, excluding T_2 and including non-drive pulleys (the tail pulley is referred to the upper branch), $T_D=2000\times9.8\times0.025\times(9.1+23)+3250=19$ kN. Upper branch tension $T_H=2000\times9.8\times0.025\times(30+23+309)+309\times9.8\times30+2\times750=269.7$ kN. To take into account additional resistances,

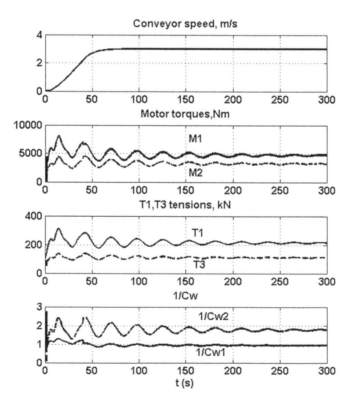

FIGURE 3.56   Start of the conveyor with two drive pulley and elastic belt with $n=8$.

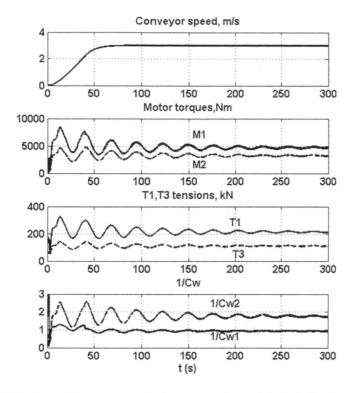

FIGURE 3.57   Start of the conveyor with two drive pulley and elastic belt with $n = 16$.

in accordance with the recommendations [14], the total tension is increased by 5%. Thus, the effective tension $T_e = (19 + 269.7) \times 1.05 = 303$ kN. If to take $T_2 = 60$ kN, then the maximum tension T_1 under the absence of a booster would be 363 kN. The required static power of all electric motors should be $303 \times 3/0.95 = 957$ kW. The same IMs, as for the previous model, are taken: for the head pulleys and booster pulleys, it is three IM with a power of 250 kW each one, they are simulated as one IM 500 kW for the main pulley and IM with power of 250 kW for the second pulley. The tension before booster pulleys is $T_{1a} = 60 + 19 + 269.7/2 = 214$ kN. Since it was taken that the booster should produce 40% of the total demanded power, it must create the tension $0.4 \times 303 = 121$ kN. Then the tension after booster is $T_{2a} = 93$ kN. The tension before the head pulleys is $T_{1b} = 93 + 269.7/2 = 228$ kN. Therefore, employment of the booster device gives an opportunity to reduce the maximum belt tension from 349 to ~228 kN (these reasoning without taking into account the additional resistances).

The conveyor developed in this way is simulated in the **Conv3** model. The drive of the second pulley of the head station and drives of the booster pulleys are modeled as library models **AC4** in the torque control mode. To drive the first pulley of the head station, the **AC4** model could also be used, in which a multiplier would be added to the circuit between the speed controller output and the torque setting for DTC, the second input of which is designated SP1; to make this change, it would be necessary to break the link with the library. However, the model developed in this way in one version of MATLAB might not function in another version, therefore, a special model has been developed for this drive that contains the same subsystems as **AC4**, but has no link with the library and works in all versions The model parameters are given in the dialog boxes of the subsystems included in this model.

The load distribution between the drives is carried out in the **Load_distrib** subsystem, by reducing the differences in the circumferential speeds of the pulleys to very small values, as in the **Conv2** model; the drives of the second head pulley and the first booster pulley are synchronized separately with the drive of the first head pulley, and the drive of the second booster pulley with the drive of the first booster pulley.

Load distribution control is carried out in the **Distrib_control** subsystem, whose block diagram is shown in Figure 3.58. PI speed controller that is placed in the M1 model can be considered the conveyor speed controller. Then, the output of this controller can be considered a required total torque of all four motors T_S^*. This quantity is transformed in the torque references for drive DTC systems by multiplication by proper factors Mi_share, $i=1, 2, 3, 4$ where $i=1$ refers to the first motor of the head pulleys, $i=2$ refers to first motor of the booster, $i=3$ refers to the second motor of the head pulleys, and $i=4$ refers to the second booster drive.

Coefficients k_1 and k_2 which determine torque distribution between the sum of the motor 1 and 3 torques and the motor 2 torque are computed by formulas (3.60), but for torques and in pu: when $T_{Sr}^* = T_S^*/T_{SN}^* < b_0$ where T_{SN}^* is the rated value of the torque sum that is taken equal to 13 kN-m that corresponds to the maximum tension of 300 kN ($300 \times 0.5/11.5 = 13$, the pulley diameter is 1 m), $k_1 = 1$, $k_2 = 0$. The value of b_0 is taken equally to 0.4 (Figure 3.47). When $T_{Sr}^* > b_0$

$$k_1 = \frac{g_h \left(T_{Sr}^* - b_0 \right) + b_0 \left(1 - T_{Sr}^* \right)}{1 - b_0} \tag{3.71}$$

$$k_2 = \frac{\left(1 - g_h \right) \left(T_{Sr}^* - b_0 \right)}{1 - b_0} \tag{3.72}$$

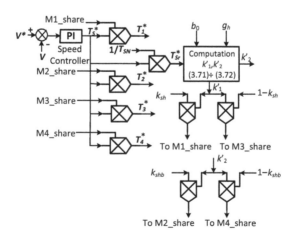

FIGURE 3.58 Block diagram of the control of the load distribution between drives in the conveyor with booster.

where g_h the fraction of the torque sum of the head station motors in the total conveyor torque. Recall that $g_h=0.6$ was taken above.

At that, $M1\_share=k_1k_{sh}$, $M3\_share=k_1(1-k_{sh})$ where k_{sh} is a fraction of the first motor in the utilization of the head station, it is taken $k_{sh}=2/3$. Analogous for the booster station, $M2\_share=k_2k_{shb}$, $M4\_share=k_2(1-k_{shb})$, it is taken $k_{shb}=2/3$. Figure 3.59 shows the realized dependences of the torques of the conveyor motors on the total load torque.

The **Conv3** model simulates conveyor speeding up. As in the previous models, the fabrication of the law of acceleration change is carried out by the serial connection of the rate limiter and the first-order lag element, the time constant of which is 5 s. The model is supplemented with the **Conveyor** subsystem, which calculates the tensions in different sections of the conveyor, as well as the coefficients C_w. The tension values are calculated taking into account the value of the acceleration a and the values of the equivalent masses of the various sections.

Tensions of the lower branch T_d, of the first half of the upper branch (subsystem **TH_1**), and of the second half of the upper branch (subsystem **TH_2**) are computed separately. After the first half, the tension decreases by the value that determines by the booster motor torques. The value $T_d+TH1+TH2-T_2$ determines the effective torque which must fabricate all conveyor motors. The scope **Scope_Cw** records values of the wrap factors for all four drive pulleys. The opportunity is provided to simulate both a fully loaded and a partially loaded conveyor by setting the values $L1$, $L2$, individually equal to zero or 1.

Some results of simulation of the loaded conveyor starting process when the speeding up time is 40 s are shown in Figures 3.60 and 3.61. It is seen that torques of the motors M1, M2, M3, M4 after speeding up are equal to 5200, 3600, 2700 and

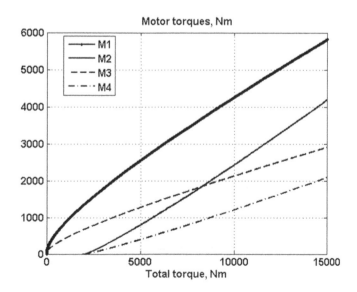

FIGURE 3.59 Realized dependences of the torques of the conveyor motors on the total load torque for the conveyor with booster.

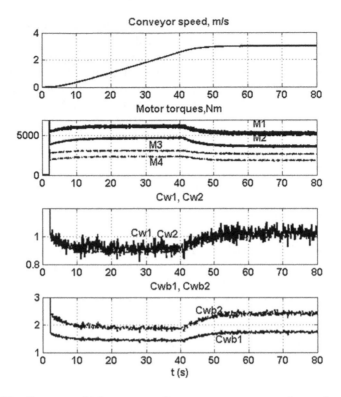

FIGURE 3.60 Conveyor with booster speeding up, motor torques and wrap factors.

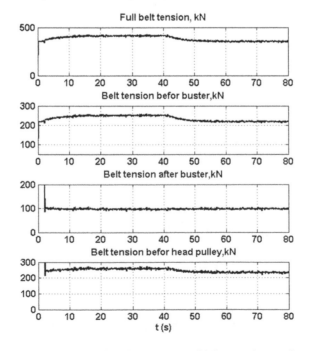

FIGURE 3.61 Tension changes when the conveyor with booster is speeding up.

1850 N-m, respectively. Thus, the torque sum is 13.35 kN-m. At this, the head station gives $7.9/13.35 = 59\%$, and booster $5.45/13.35 = 41\%$, as intended, with an error of 1%. The wrap factors are in the safe range.

It can be seen from Figure 3.61 that the total tension that is necessary to overcome all resistance torques is ~325 kN (here and below, the steady state is considered). The tension before booster is 220 kN, and after it 98 kN, a loss of tension corresponds to motor torques of 5.58 kN. In front of the head station, the tension increases to 230 kN. These data correspond to the calculated values above.

Some peculiarities appear when the conveyor is unevenly loaded, especially when the second half is loaded, but the first is not [11]. The results of modeling such a condition are shown in Figures 3.62 and 3.63 (The constant **Load**=0, constant **Load1**=1). The torques of the motors M1, M2, M3, M4 after acceleration are 3400, 1650, 1750, and 800 N-m, respectively. Thus, when the load decreases, most of the work is done by the motors of the head station, as it was intended.

It is interesting to estimate the possibility to decrease tension T_2. The plots when T_2 is reduced from 60 kN down to 40 kN are shown in Figure 3.64. It is seen that the condition of no slippage on the second booster pulley is violated.

Start conditions are greatly eased for an unloaded conveyor, which is a more frequent occurrence. Figures 3.65 and 3.66 show the plots of the same quantities as in Figures 3.60 and 3.61, but for the start of an empty conveyor. The acceleration value has been doubled. It can be seen that the start is provided by all four motors, but after it is completed, the torques of the booster motors decrease to zero, since the effective

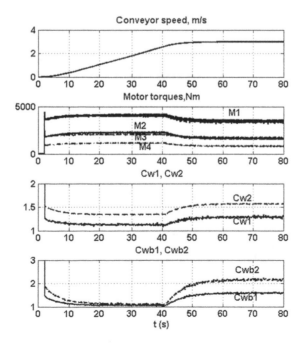

FIGURE 3.62 Conveyor with booster speeding up under uneven load, motor torques, and wrap factors.

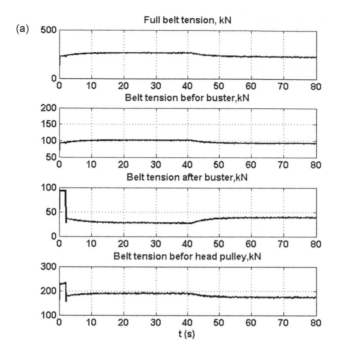

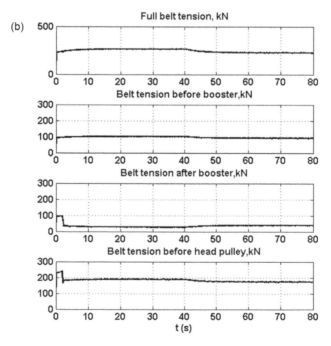

FIGURE 3.63 Tension changes when the conveyor with booster is speeding up under uneven load.

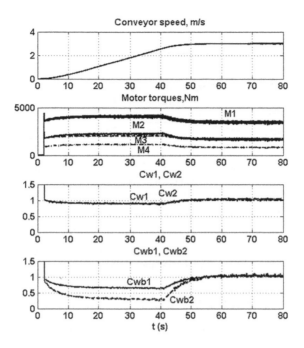

FIGURE 3.64 Conveyor with booster speeding up with reduced T_2 tension.

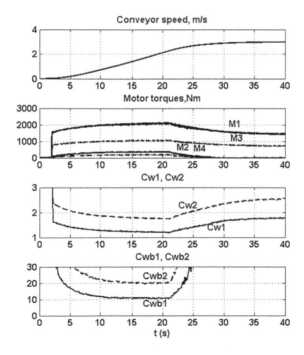

FIGURE 3.65 Unloaded conveyor with booster speeding up, motor torques, and wrap factors.

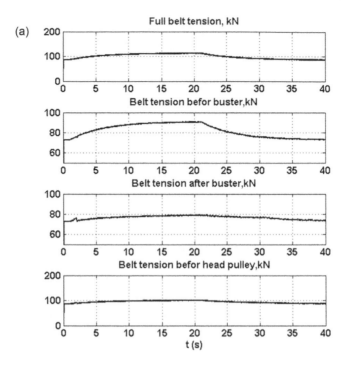

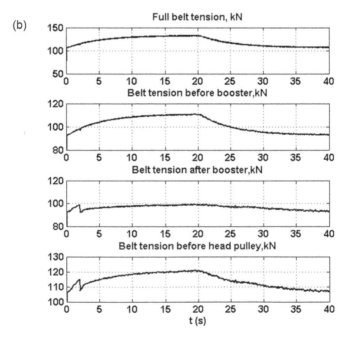

FIGURE 3.66 Tension changes when the unloaded conveyor with booster is speeding up.

tension is 50 kN (the last graph on the **Scope_tens** oscilloscope), and the required total torque is $50 \times 0.5/(11.5 \times 0.95) = 2.3$ kN-m, at which the booster torques are equal to zero (Figure 3.59). Wrap factors exceed the minimum allowable values by a large margin.

REFERENCES

1. ABB. KATALOG JUNI 2018, Niederspannung Motoren für die Prozessindustrie 400 V 50 Hz, 460V 60 Hz
2. Krause, P. C., Wasynczuk, O., Sudhoff, S. D. *Analysis of Electric Machinery.* IEEE Press, Piscataway, NJ, 2002.
3. Kress, R. L., Jansen, J. F., Noakes, M. W. Experimental Implementation of a Robust Damped-Oscillation Control Algorithm on a Full-Sized, Two-Degree-of-Freedom, AC Induction Motor-Driven Crane. ISRAM '94, Fifth International Symposium on Robotics and Manufacturing, Maui, Hawaii, August 14–17, 1994.
4. Santhi, L. R., M, L. B. Position Control and Anti-Swing Control of Overhead Crane Using LQR. *International Journal of Scientific Engineering and Research (IJSER),* Vol. 3, No. 8, August 2015.
5. Abdel-Rahman, E. M., Nayfeh, A. H., Masoud, Z. N. Dynamics and Control of Cranes: A Review. *Journal of Vibration and Control,* Vol. 9, No. 7, July 2003.
6. Alzinger, E., Brozovic, V. Automation and Control System for Grab Cranes. Brown Boveri Rev. No. 9/10, 1983.
7. Lobov, N. A. *Dynamics of Hoisting Cranes (in Russian).* Mashinostroenie, Moscow, 1987.
8. Bose, B. K. *Modern Power Electronics and AC Drives.* Prentice Hall PTR, Upper Saddle River, NI, 2002.
9. Stepanov, A. G. *Dynamics of Machines.* Yekaterinburg, Ural Branch of the Russian Academy of Sciences, 1999.
10. Grishko A. P. Stationary machines. Volume 1. Mine *Lifting Installations.* M. Publishing House of the Moscow State Mining University, Moscow, 2006.
11. Churchill, F., T. Dynamic Load Sharing for Conveyor Belts with Multiple Drive Stations. 12th WVU International Mining Electrotechnology Conference, Morgantown, WV, USA, 1994.
12. Holtom, A., Keller, D., Winkle, P. V. Drives for Overland Conveyors and Lessons Learned. IEEE-IAS/PCA Cement Industry Technical Conference, National Harbor, MD, USA, 2014.
13. CEMA. Engineering Conference of the Conveyor Equipment Manufacturers Association, *Belt Conveyors for Bulk Materials.* Fifth Edition, PDF Version, Conveyor Equipment Manufacturers Association, United States of America, July, 2002.
14. DIN 22101. Continuous conveyors. Belt conveyors for loose bulk materials. Basis for calculation and dimensioning. English translation of DIN 22101:2011–12.
15. Galkin V. I., Dmitriev V. G., Dyachenko V. P., Zapenin I. V., Sheshko E. E. *Modern Theory of Belt Conveyors of Mining Enterprises.* Publishing House of the Moscow State Mining University, Moscow, 2005.
16. Lodewijks, G. Dynamics of Belt Systems, Ph.D. thesis, Delft University of Technology, 1996.
17. Lodewijks, G. Two Decades Dynamics of Belt Conveyor Systems. *Bulk Solids Handling,* Vol. 22, No. 2, 2002.
18. He, D., Pang, Y., Lodewijks, G. Determination of Acceleration for Belt Conveyor Speed Control in Transient Operation. *IACSIT International Journal of Engineering and Technology,* Vol. 8, No. 3, March 2016.

List of the Appended Models and Programs

Rev-torque:	Draughting schedule calculation
Blm1.m:	Rolling force in blooming mill calculation
Blum1:	Simple Blooming, 13 Passes
Blum2:	Individual electric drive of rolls with DC motors 6800 kW, 60/90 rpm
Blum2a:	Individual electric drive of rolls with DC motors 6800 kW, 60/90 rpm, with mill consideration
Blum 2b:	Individual electric drive of rolls with DC motors 6800 kW, 60/90 rpm, elastic coupling
Blum 3:	Individual electric drive of rolls with synchronous motors 8000 kVA
Blum 3a:	Individual electric drive of rolls with synchronous motors 8000 kVA, with mill consideration, with active rectifier
Blum 3a1:	Individual electric drive of rolls with synchronous motors 8000 kVA, with mill consideration, without active rectifier
DTC_pos:	Screwdown drive
Blum 4:	Roll and Screwdown Drives
Plt1.m:	Rolling force in plate mill calculation
Plate1:	Simple Plate Mill, 13 Passes
Plate2:	Plate Mill, Roll, and Screwdown Drives, simple Model
Plate1a:	Simple Plate Mill, 13 Passes, Thickness Control
Plate3:	Plate Mill Drive for one Roll
Plate3a:	Plate Mill, Top Roll, elastic Coupling Consideration
Plate32:	Plate Mill Drive, one Roll, with Active Rectifiers and Individual Transformers
Plate 33:	Plate Mill Drive, one Roll, with Active Rectifiers and Common Transformer
Plate34:	Plate Mill Drive, one Roll, with Active Rectifiers, Common DC Bus
Mill_rough1:	Two stands of the roughing group, simplified model

Mill_rough2:	Three stands of the roughing group, simplified model
Mill_rough3:	Three stands of the roughing group, complete model, without front rectifiers
Mill_rough2a:	Three stands of the roughing group, simplified model, strip output
Mill_rough3a:	Three stands of the roughing group, complete model
Drive_12000_6Ph:	Electrical drive with 6-phase synchronous motor
Looper2:	Simple Looper Model
Looper2a:	Simple Model of the Stands 3&4 with Looper
P_calc:	Thickness control
P_calc1:	Thickness control in two stands
Looper2b:	Stands 3 and 4 Acceleration
Looper2c:	Simple Model of the Stands 3&4 with Looper and Thickness Control
Mill1_7:	Hot Strip Finishing Mill
Thickness:	Comparison of the methods of thickness control
Mill1_7b:	Hot Strip Finishing Mill, with thickness control
Mill1_7c:	Hot Strip Finishing Mil with direct Tension Measurement
Mill1_7d:	Hot Strip Finishing Mil with direct Tension Measurement and Thickness Control
Mill_finish2_3:	Finishing Stands 2&3
Mill_finish4_5:	Finishing Stands 4&5
Mill_finish6_7:	Finishing Stands 6&7
Finish_Mill_Drive_12MW_60:	Finish Mill Drive, 12 MW, 60 rpm
Finish_Mill_Drive_12MW_150:	Finish Mill Drive 12 MW, 150 rpm
Finish_Mill_Drive_12MW_250:	Finish Mill Drive 12 MW, 250 rpm
Finish_Mill_Drive_10MW_300:	Finish Mill Drive 10 MW, 300 rpm
Finish_Mill_Drive_8MW_400:	Finish Mill Drive 8 MW, 400 rpm
DTC_shear1:	Flying Shear with 3-level VSI and DTC, active Rectifier
DTC_shear2:	Flying Shear with 3-level VSI and DTC, cutting for given length
Reel1:	Reel with indirect tension control
Reel1a:	Reel with indirect tension control, with active rectifier
Reel2:	Reel with direct tension control
Reel2a:	Reel with direct tension control, with active rectifier
Reel3:	Uncoiler with indirect tension control

Paper2e:	Turning and couch rolls, detailed model
Paper3:	Two-roll press with one drive with DTC
Paper3b:	Drive with PMSM
Paper3a:	Two-Roll Press with two PMSM Drives, with controlled DC Source, with position sensors
Paper3a2:	Two-Roll Press with two PMSM Drives, with Controlled DC Source, without sensors
Paper3c:	Three-Roll Press with two PMSM Drives, with Controlled DC Source, with position sensor
Paper3c2:	Three-Roll Press with two PMSM Drives, with Controlled DC Source, without sensors
Paper4:	Dryer section with DTC
Paper4_1:	Dryer section with DTC, fast steady-state
Paper4a:	Two last drying sections
Paper5:	Calender drive with DTC
Paper5a:	Dryer and Calender
Paper6:	Reel&Calender
Paper6a:	Reel separately
Paper6b:	Reel with tension control
Paper6b1:	Reel with tension control, speed up process
Paper7:	Supercalender, main drive
Paper7a:	Unwinder of the supercalender separately
Paper7b:	Reel of the supercalender separately
Paper7c1:	Supercalender Electric Drives, with active rectifiers
Paper7c3:	Supercalender Electric Drives, winding cycle termination
Paper8a:	Slitter unwinder separately
Paper8a1:	Slitter unwinder separately_1
Paper8b:	Two-drum winder
Paper8c1:	Slitter drives, with active Rectifier
Paper8c2:	Slitter, winding cycle termination
Crane1a:	Hoisting drive with an active rectifier
Crane1b:	Trolley drive with common VSI, first option
Crane2b:	Trolley drive with common VSI, second option
Crane1c:	Traveling crane electrical drives
Crane1d:	Traveling crane, simple model, lifting and moving in turn
Crane1d4:	Traveling crane, simple model, lifting and moving simultaneously
Crane1d1:	Traveling crane, simple model, with double notch filter
Crane1d2:	Traveling crane, simple model, two-stage speeding up

Index

Note: **Bold** page numbers refer to tables and *italic* page numbers refer to figures.

Milton Keynes UK
Ingram Content Group UK Ltd.
UKHW031129141024
449569UK00006B/318